Ergebnisse der Mathematik und ihrer Grenzgebiete

Band 73

J. Milnor · D. Husemoller

Symmetric Bilinear Forms

Springer-Verlag New York Heidelberg Berlin
1973

John Milnor
The Institute for Advanced Study, Princeton, New Jersey 08540, U.S.A.

Dale Husemoller
Haverford College, Haverford, Pennsylvania 19041, U.S.A.

AMS Subject Classifications (1970):
Primary 10 C 05 · Secondary 10 E 25, 15 A 63

ISBN 0-387-06009-X Springer-Verlag New York Heidelberg Berlin
ISBN 3-540-06009-X Springer-Verlag Berlin Heidelberg New York

Typesetting, printing and binding: Universitätsdruckerei H. Stürtz AG, Würzburg.

Preface

The theory of quadratic forms and the intimately related theory of symmetric bilinear forms have a long and rich history, highlighted by the work of Legendre, Gauss, Minkowski, and Hasse. (Compare [Dickson] and [Bourbaki, **24**, p. 185].) Our exposition will concentrate on the relatively recent developments which begin with and are inspired by Witt's 1937 paper "Theorie der quadratischen Formen in beliebigen Körpern." We will be particularly interested in the work of A. Pfister and M. Knebusch. However, some older material will be described, particularly in Chapter II. The presentation is based on lectures by Milnor at the Institute for Advanced Study, and at Haverford College under the Phillips Lecture Program, during the Fall of 1970, as well as lectures at Princeton University in 1966. We want to thank J. Cunningham, M. Knebusch, M. Kneser, A. Rosenberg, W. Scharlau and J.-P. Serre for helpful suggestions and corrections.

Prerequisites. The reader should be familiar with the rudiments of algebra, including for example the concept of tensor product for modules over a commutative ring. A few individual sections will require quite a bit more.

The logical relationship between the various chapters can be roughly described by the diagram below. There are also five appendices, largely self-contained, which treat special topics.

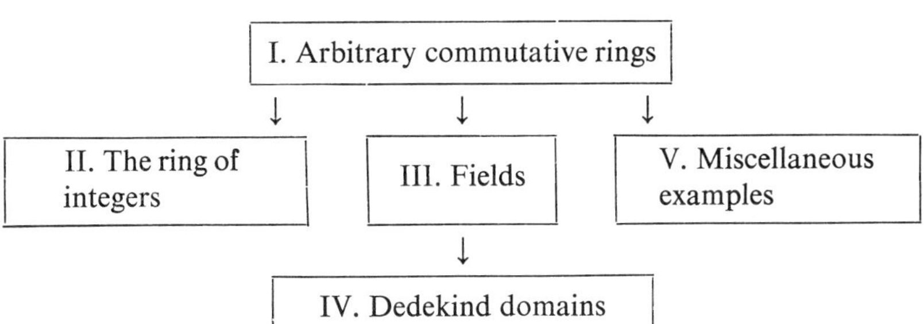

Contents

Chapter I. Basic Concepts

This chapter will define the concept of an *inner product space* over a commutative ring R, and describe basic constructions which are independent of the ring R. In particular it introduces the *Witt ring* $W(R)$, which will play a central role in later chapters. Roughly speaking, $W(R)$ is the collection of all symmetric inner product spaces X over R modulo the collection of "split" inner product spaces. The inner product space X is said to be *split* if $X = X_1 + X_2$ where the submodules X_1 and X_2 are dually paired by the inner product, and $X_1 \cdot X_1 = 0$.

§ 1. Bilinear Forms and Inner Products

Let R be a commutative ring with 1, and let X be a left R-module.

(1.1) Definition. A *bilinear form* on X is a function

$$\beta\colon X \times X \to R$$

such that $\beta(x, y)$ is R-linear as a function of x for fixed y, and R-linear as a function of y for fixed x. Such a bilinear form β will be called an *inner product* on X if the following strong non-degeneracy conditions are satisfied. For each R-linear map

$$\varphi\colon X \to R$$

there should exist one and only one element x_0 in X so that the homomorphism

$$y \mapsto \beta(x_0, y)$$

from X to R is equal to φ. Furthermore there should exist one and only one $y_0 \in X$ so that the homomorphism

$$x \mapsto \beta(x, y_0)$$

is equal to φ. In other words the two homomorphisms

$$x_0 \mapsto \beta(x_0, \ ), \qquad y_0 \mapsto \beta(\ , y_0)$$

from X to the dual module $\mathrm{Hom}_R(X, R)$ should be bijective.

The notation $\beta(x, y) = x \cdot y$ will usually be used for an inner product.

If β is a bilinear form or inner product on X, then the pair (X, β) is called a *bilinear form module* or an *inner product module* over R. Two bilinear form modules (X, β) and (X', β') are *isomorphic* if there is an R-linear bijection $f: X \to X'$ satisfying $\beta'(f(x), f(y)) = \beta(x, y)$ for all x and y in X.

By abuse of language, we will use the abbreviated notation X for a bilinear form module, whenever there is no danger of confusion.

We will be particularly interested in R-modules which are finitely generated and projective[1].

(1.2) Definition. An inner product module (X, β) will be called an inner product *space* if X is finitely generated and projective over R.

Also, we will be particularly interested in inner products which are symmetric.

(1.3) Definition. A bilinear form or inner product β is *symmetric* if

$$\beta(x, y) = \beta(y, x)$$

for all x and y. Similarly it is *skew-symmetric* if

$$\beta(x, y) = -\beta(y, x)$$

for all x and y, and *symplectic* (or alternating) if

$$\beta(x, x) = 0$$

for all x. The computation $\beta(x, y) + \beta(y, x) = \beta(x+y, x+y) - \beta(x, x) - \beta(y, y)$ shows that every symplectic inner product is skew-symmetric. Over a ring R where 2 is a unit, the converse holds, since the equation $\beta(x, x) = -\beta(x, x)$ then implies that $\beta(x, x) = 0$.

Two elements x and y of a bilinear form module are called *orthogonal* if $\beta(x, y) = 0$. In making use of the concept of orthogonality, we will always assume that β is either symmetric or skew-symmetric, so that $\beta(x, y) = 0$ implies $\beta(y, x) = 0$. Compare [Artin, p. 110].

If X is an inner product module, note that an element x is orthogonal to every $y \in X$ if and only if $x = 0$. Over a field the converse statement is true: If (X, β) is a bilinear form space over a field such that only the zero vector in X is orthogonal to every $y \in X$, then clearly β is an inner product on X.

---

[1] A module P is *projective* if there exist a module Q so that $P \oplus Q$ is free (i.e., isomorphic to a direct sum of copies of R).

§ 2. Bilinear Forms over a Free Module

If X is a finitely generated free R-module with basis $e_1, \ldots, e_n$, then the integer n is called the *rank* or *dimension* of X, denoted $\operatorname{rk}(X)$. Since R is commutative, the rank is uniquely defined.

If X has basis $e_1, \ldots, e_n$, then any bilinear form β on X gives rise to an $n \times n$ matrix $B = (\beta_{ij})$ where

$$\beta_{ij} = \beta(e_i, e_j).$$

This matrix determines the bilinear form uniquely, since if $x = \sum \xi_i e_i$ and $y = \sum \eta_j e_j$ then

$$\beta(x, y) = \sum \beta_{ij} \xi_i \eta_j.$$

(2.1) Definition. Given any $n \times n$ matrix $B = (\beta_{ij})$ with entries in R, the symbol

$$\langle B \rangle = \langle B \rangle_R$$

will stand for the free bilinear form space over R with basis $e_1, \ldots, e_n$ and with bilinear form $\beta(e_i, e_j) = \beta_{ij}$.

(2.2) Lemma. *This bilinear form is an inner product if and only if the matrix B is invertible (i.e., has a 2-sided inverse).*

This is clear since the homomorphism $x \mapsto \beta(x, \ )$ from X to the dual module $\operatorname{Hom}_R(X, R)$, equipped with the dual basis $e_1^*, \ldots, e_n^*$, is given by $e_i \mapsto \sum_j \beta_{ij} e_j^*$. □

Note that the bilinear form space $\langle B \rangle$ is symmetric if and only if $B = B^t$, skew-symmetric if and only if $B^t = -B$, and symplectic if and only if the matrix B has zeros along the diagonal and satisfies $B^t = -B$. (Here B^t stands, of course, for the transpose of the matrix B.)

Now let us see what happens if we change the basis.

(2.3) Lemma. *The bilinear form space $\langle B \rangle$ is isomorphic to $\langle B' \rangle$ if and only if*

$$B' = A B A^t$$

for some invertible $n \times n$ matrix A.

For if $e'_1, \ldots, e'_n$ is a new basis, then

$$e'_i = \alpha_{i1} e_1 + \cdots + \alpha_{in} e_n$$

for some invertible matrix (α_{ik}), and it follows that

$$B(e'_i, e'_j) = \sum \alpha_{ik} \beta_{kl} \alpha_{jl},$$

summed over $1 \leqq k \leqq n$, $1 \leqq l \leqq n$. □

The following special case is of particular interest.

(2.4) Example. Let u be any element in the group $R^\bullet$ consisting of all units in R. Then the symbol $\langle u \rangle$ denotes the symmetric inner product space having one basis element e_1, where $e_1 \cdot e_1 = u$. Note that

$$\langle u \rangle \cong \langle u' \rangle$$

if and only if $u' = \alpha^2 u$ for some $\alpha \in R^\bullet$.

A useful invariant of free inner product spaces is the determinant. Let $R^{\bullet 2}$ denote the subgroup of $R^\bullet$ consisting of all squares of units.

(2.5) Definition. The *determinant* of a free inner product space X is the element of the quotient group $R^\bullet / R^{\bullet 2}$ represented by $\det(B)$, where B is any matrix with $\langle B \rangle \cong X$.

More generally, if X is a free bilinear form module, then $\det(X)$ is the element of the quotient monoid $R/R^{\bullet 2}$ represented by $\det(B)$, where B is any matrix with $\langle B \rangle \cong X$. It follows from (2.3) that this determinant is well defined.

We conclude with one more useful and classical construction. Given a basis $e_1, \ldots, e_n$ for a free inner product space X, the *dual basis* $e_1^\#, \ldots, e_n^\#$ for X is defined by the conditions

$$e_i \cdot e_k^\# = 0 \quad \text{for } i \neq k,$$

and

$$e_i \cdot e_i^\# = 1.$$

(2.6) Lemma. *To each basis for a free inner product space there corresponds a unique dual basis.*

For the matrix $(\beta_{ij}) = (e_i \cdot e_j)$ is invertible, with inverse matrix (γ_{jk}). The equations

$$e_k^\# = \gamma_{1k} e_1 + \cdots + \gamma_{nk} e_n$$

now yield the required dual basis. $\square$

§ 3. Orthogonal Sums

Let $X_1, \ldots, X_n$ be bilinear form modules, with bilinear forms $\beta_1, \ldots, \beta_n$ respectively. The *orthogonal sum* $X_1 \oplus \cdots \oplus X_n$ is defined to be the direct sum of the modules X_i with bilinear form β defined by the equation

$$\beta(x_1 \oplus \cdots \oplus x_n, y_1 \oplus \cdots \oplus y_n) = \sum \beta_i(x_i, y_i)$$

summed over $1 \leqq i \leqq n$.

Evidently $X_1 \oplus \cdots \oplus X_n$ is an inner product module (or an inner product space) if and only if each X_i is an inner product module (or an inner product space). If the X_i are free and finitely generated, note that

$$\operatorname{rk}(X_1 \oplus \cdots \oplus X_n) = \sum \operatorname{rk}(X_i),$$

and

$$\det(X_1 \oplus \cdots \oplus X_n) = \prod \det(X_i).$$

The following lemma is easy to prove, but extremely important. Let X be a bilinear form module, and M a submodule. *We assume that the bilinear form β is either symmetric or skew-symmetric, so that $\beta(x, y) = 0$ implies $\beta(y, x) = 0$.*

(3.1) Orthogonal decomposition lemma. *If the bilinear form β restricted to $M \times M$ is an inner product on M, then X is isomorphic to the orthogonal sum $M \oplus M^\perp$.*

Here $M^\perp$ denotes the *orthogonal complement*, consisting of all $x \in X$ such that $\beta(x, M) = 0$.

Proof. If $m \in M \cap M^\perp$, then $\beta(m, m') = 0$ for all $m' \in M$, and therefore $m = 0$. Thus to prove (3.1) it suffices to show that every x in X can be written as a sum $m + y$ with $m \in M$ and $y \in M^\perp$.

Given $x \in X$, consider the linear form $m' \mapsto \beta(x, m')$ on M. By the definition of inner product, there exists one and only one element $m \in M$ so that

$$\beta(m, m') = \beta(x, m')$$

for all m'. Then $x - m \in M^\perp$, and we have

$$x = m + (x - m)$$

as required. This completes the proof. ∎

(3.2) Theorem. *Let X be a symmetric or skew-symmetric bilinear form module, and let $x_1, \ldots, x_k$ be elements such that the $k \times k$ matrix $(\beta(x_i, x_j))$ is invertible. Then $x_1, \ldots, x_k$ are linearly independent, and*

$$X \cong M \oplus M^\perp$$

where M denotes the free module spanned by the x_i.

Proof. Since any relation $\rho_1 x_1 + \cdots + \rho_k x_k = 0$ would contradict the hypothesis that $(\beta(x_i, x_j))$ is invertible, this follows easily from the lemma. ∎

This theorem has many applications. Here are some examples.

(3.3) Corollary. *If X is a finitely generated symmetric bilinear form module, then*

$$X \cong \langle u_1 \rangle \oplus \cdots \oplus \langle u_k \rangle \oplus N$$

where $u_1, \ldots, u_k$ are units, and $\beta(x, x)$ is a non-unit for every $x \in N$.

For if X contains some element x_1 such that $\beta(x_1, x_1) = u_1$ is a unit, then

$$X \cong (R\,x_1) \oplus (R\,x_1)^{\perp}$$

by (3.2), where the submodule

$$R\,x_1 \cong \langle u_1 \rangle$$

is free. Now apply the same construction to $(R\,x_1)^{\perp}$, and continue inductively.

This procedure must terminate after finitely many steps. For suppose that the module X is generated by n elements. If the construction continued for more than n steps, we could construct a homomorphism from a free module of rank n onto a free module of rank $n+1$. Since R is commutative, this is impossible; and this completes the proof. □

If R is a field, it follows that $\beta(x, x) = 0$ for every $x \in N$, so that N is symplectic. In fact if R is a field of characteristic $\neq 2$, then β restricted to $N \times N$ being both symmetric and symplectic must actually be zero. In the case of an inner product, this implies that N itself must be zero. More generally consider a *local ring* (i.e., a ring with unique maximal ideal).

(3.4) Corollary. *If R is a local ring in which 2 is a unit, then every symmetric inner product space X over R possesses an orthogonal basis.*

That is, X possesses a basis $e_1, \ldots, e_k$ so that $e_i \cdot e_j = 0$ for $i \neq j$. In other words

$$X \cong \langle u_1 \rangle \oplus \cdots \oplus \langle u_k \rangle$$

for suitable units $u_1, \ldots, u_k$.

Proof of (3.4). Consider the submodule N of (3.3). As an orthogonal summand of an inner product space, N must itself be an inner product space. Suppose that N were non-zero. Since every finitely generated projective over a local ring is free, see [Swan, 1968] or [Milnor, Intr. algebr. K-theory], we could choose a basis $e_1, \ldots, e_n$ for N, with $n \geqq 1$. Let $e_1^{\#}, \ldots, e_n^{\#}$ be the dual basis. Then the computation

$$2 = 2\,e_1 \cdot e_1^{\#} = (e_1 + e_1^{\#}) \cdot (e_1 + e_1^{\#}) - e_1 \cdot e_1 - e_1^{\#} \cdot e_1^{\#}$$

would show that 2 belonged to the ideal of non units, contradicting our hypothesis. Thus $N = 0$, which completes the proof. □

Here is a final example. Let X be a symplectic inner product space. By a *symplectic basis* for X we will mean a basis $e_1, \ldots, e_n$ such that the associated inner product matrix $(e_i \cdot e_j)$ has the form $\begin{pmatrix} 0 & I \\ -I & 0 \end{pmatrix}$.

(3.5) Corollary. *If R is either a Dedekind domain* (see p. 91 for a definition) *or a local ring, then every symplectic inner product space over R is free, and possesses a symplectic basis.*

Thus the rank of such a space is always even, and the determinant is always the identity element of $R^\bullet/R^{\bullet 2}$. (More generally, for any free symplectic bilinear form space, the determinant in $R/R^{\bullet 2}$ has a canonical square root in $R/R^\bullet$ called the "Pfaffian". See [Bourbaki, v. **24**, p. 83].)

Proof. We must first construct two elements x_1 and x_2 in X so that $x_1 \cdot x_2 = 1$. If R is a local ring, then the projective module X is necessarily free, so choosing a basis $e_1, \ldots, e_n$ and a dual basis $e_1^\#, \ldots, e_n^\#$, the two vectors e_1 and $e_1^\#$ will serve.

In the case of a Dedekind ring, a classical theorem of Steinitz[2] asserts that the projective module X is the direct sum of a free module with basis $e_1, \ldots, e_n$ and an ideal $\mathfrak{a} \subset R$. If $n \geqq 1$, then we can again choose $e_1^\#$ so that $e_1 \cdot e_1^\# = 1$. But if $n = 0$, then $X \cong \mathfrak{a}$, and a bilinear form $\beta\colon \mathfrak{a} \times \mathfrak{a} \to R$ which is symplectic must clearly be zero. Hence the case $X \cong \mathfrak{a} \neq 0$ cannot occur.

Thus if $X \neq 0$ there exist elements x_1 and x_2 with $x_1 \cdot x_2 = 1$. The 2×2 matrix

$$(x_i \cdot x_j) = \begin{pmatrix} 0 & 1 \\ -1 & 0 \end{pmatrix}$$

is evidently invertible, so by (3.2) these two elements span a free orthogonal summand. An easy inductive argument now completes the proof. □

An example of a symplectic inner product space with no symplectic basis will be constructed in Chapter V, § 2.

§ 4. Witt's Theorem

Let X be a symmetric bilinear form module over the ring R. Suppose that we are given an orthogonal sum decomposition $X = M \oplus N$.

(4.1) Definition. The *reflection* of X with respect to (M, N) is the linear transformation $r\colon X \to X$ which leaves M pointwise fixed and carries each point of N to its negative.

---

[2] See for example [Milnor, Introduction to Algebraic K-Theory, pp. 10–11].

Thus r maps each sum $x=m+n$ in X to $r(x)=m-n$. Evidently r is an involution

$$r(r(x))=x,$$

and evidently r preserves the bilinear form,

$$\beta(r(x), r(y))=\beta(x, y)$$

for all x and y. If 2 is a unit in R, then conversely it is easy to show that every linear involution preserving the bilinear form on X is a reflection.

(4.2) Lemma. *Suppose that R is a local ring in which 2 is a unit. If x and y are elements in the symmetric bilinear form module X such that*

$$\beta(x, x)=\beta(y, y)$$

is a unit of R, then there exists a reflection of X carrying x to y.

Proof. Express x as the sum of two mutually orthogonal vectors $u=(x+y)/2$ and $v=(x-y)/2$. Then

$$\beta(x, x)=\beta(u, u)+\beta(v, v).$$

Since R is local, at least one of the two ring elements $\beta(u, u)$ and $\beta(v, v)$ must be a unit. If $\beta(u, u)$ is a unit, then $X=(Ru)\oplus(Ru)^{\perp}$, and the reflection with respect to $((Ru), (Ru)^{\perp})$ carries $u+v=x$ to $u-v=y$. Similarly, if $\beta(v, v)$ is a unit, then reflection with respect to $((Rv)^{\perp}, (Rv))$ carries x to y. This completes the proof. ∎

(4.3) Corollary. *With R as above, if X is a symmetric inner product space of rank n over R, then every automorphism f of X can be expressed as the composition of n reflections.*

Proof by induction. By (3.4) there exists an orthogonal basis $e_1, \ldots, e_n$ for X. Choose a reflection r_1 carrying $f(e_1)$ to e_1. Then $r_1 f$ fixes e_1 and hence carries the space $(Re_1)^{\perp}$ of rank $n-1$ to itself. Therefore $r_1 f$ restricted to $(Re_1)^{\perp}$ is a composition $r_2 \ldots r_n$ of reflections. Extending each r_i to X by setting $r_i(e_1)=e_1$ for $i>1$, we have $f=r_1 \ldots r_n$ as required. ∎

Another corollary, more important for our purposes, is the following. We continue to assume that R is a local ring in which 2 is a unit.

(4.4) Witt's theorem. *Let X, Y, Z be inner product spaces over R. If $X\oplus Y\cong X\oplus Z$, then $Y\cong Z$.*

Proof. Since X is an orthogonal sum of rank 1 spaces by (3.4), it suffices to prove this theorem when X is free of rank 1. Let e be a basis element for X, and let

$$f\colon X\oplus Y\to X\oplus Z$$

be an arbitrary isomorphism. To avoid confusion, let 0_X, 0_Y, and 0_Z denote the zero elements in X, Y and Z respectively. Then the two elements $f(e \oplus 0_Y)$ and $e \oplus 0_Z$ of $X \oplus Z$ satisfy the hypothesis of (4.2), so there exists a reflection r of $X \oplus Z$ carrying $f(e \oplus 0_Y)$ to $e \oplus 0_Z$. Now the isomorphism

$$rf\colon X \oplus Y \to X \oplus Z$$

carries $e \oplus 0_Y$ to $e \oplus 0_Z$, and hence carries the orthogonal complement $0_X \oplus Y$ isomorphically to $0_X \oplus Z$. This completes the proof. □

Note that Witt's theorem is definitely false when 2 is a non-unit in R. To prove this, start with the isomorphism

$$\langle -1 \rangle \oplus \langle -1 \rangle \oplus \langle 1 \rangle \cong \langle -1 \rangle \oplus H,$$

where H denotes the *hyperbolic plane*, free of rank 2 with inner product matrix $\begin{pmatrix} 0 & 1 \\ 1 & 0 \end{pmatrix}$. In fact if e_1, e_2, e_3 are orthogonal vectors with $e_1 \cdot e_1 = e_2 \cdot e_2 = -1$ and $e_3 \cdot e_3 = 1$, then the three vectors

$$e_1 + e_2 + e_3, \qquad e_1 + e_3, \qquad e_2 + e_3$$

form a new basis, with inner product matrix

$$\begin{pmatrix} -1 & 0 & 0 \\ 0 & 0 & 1 \\ 0 & 1 & 0 \end{pmatrix}.$$

But if 2 is a non-unit, then

$$\langle -1 \rangle \oplus \langle 1 \rangle \not\cong H,$$

since every $h \in H$ clearly satisfies

$$h \cdot h \equiv 0 \mod 2R.$$

(Compare the discussion in Chapter V, § 1.)

It is interesting to note that Witt's theorem remains true for *quadratic forms* over a field of characteristic 2; see Appendix 1.

§ 5. Tensor Products and Exterior Powers

Let $X_1, \ldots, X_n$ be bilinear form modules over R, with bilinear forms $\beta_1, \ldots, \beta_n$ respectively. Then the tensor product $X_1 \otimes \cdots \otimes X_n$ over R can be made into a bilinear form module as follows.

(5.1) Lemma. *There is one and only one bilinear form β on $X_1 \otimes \cdots \otimes X_n$ which satisfies the identity.*

$$\beta(x_1 \otimes \cdots \otimes x_n, y_1 \otimes \cdots \otimes y_n) = \prod_{i=1}^{n} \beta_i(x_i, y_i)$$

for all x_i and y_i in X_i, $1 \leqq i \leqq n$.

Proof. The $2n$-linear function

$$(x_1, \ldots, x_n, y_1, \ldots, y_n) \mapsto \beta_1(x_1, y_1) \ldots \beta_n(x_n, y_n)$$

from $X_1 \times \cdots \times X_n \times X_1 \times \cdots \times X_n$ to R gives rise to an associated linear function

$$X_1 \otimes \cdots \otimes X_n \otimes X_1 \otimes \cdots \otimes X_n \to R.$$

Composing with the canonical bilinear function

$$(X_1 \otimes \cdots \otimes X_n) \times (X_1 \otimes \cdots \otimes X_n) \to X_1 \otimes \cdots \otimes X_n \otimes X_1 \otimes \cdots \otimes X_n,$$

we obtain the required bilinear form β. □

(5.2) Remark. If each X_i is symmetric, then the tensor product is clearly symmetric. More generally, if each X_i is ε_i-symmetric (where 1-symmetric means symmetric and (-1)-symmetric means skew-symmetric), then the tensor product $X_1 \otimes \cdots \otimes X_n$ is $(\varepsilon_1 \ldots \varepsilon_n)$-symmetric. Similarly, if X_1 is symmetric and X_2 is symplectic, then $X_1 \otimes X_2$ is symplectic.

Now let us suppose that each X_i is an inner product space.

(5.3) Lemma. *If $X_1, \ldots, X_n$ are inner product spaces over R, then $X_1 \otimes \cdots \otimes X_n$ is an inner product space over R.*

Proof. Since each module X_i is a direct summand of a finitely generated free R-module, it follows easily that the tensor product $X_1 \otimes \cdots \otimes X_n$ is a direct summand of a finitely generated free R-module.

Let X^* denote the module $\mathrm{Hom}(X, R)$ dual to X. Each inner product β_i gives rise to an associated bijection

$$\bar{\beta}_i \colon X_i \to X_i^*,$$

where $\bar{\beta}_i(x)(y) = \beta_i(x, y)$. The homomorphism

$$\bar{\beta} \colon X_1 \otimes \cdots \otimes X_n \to (X_1 \otimes \cdots \otimes X_n)^*$$

associated with our bilinear form β, can be expressed as a composition

$$\bar{\beta} = \eta \circ (\bar{\beta}_1 \otimes \cdots \otimes \bar{\beta}_n)$$

where η is the homomorphism from $X_1^* \otimes \cdots \otimes X_n^*$ to $(X_1 \otimes \cdots \otimes X_n)^*$ which maps each generator $f_1 \otimes \cdots \otimes f_n$ to the homomorphism $x_1 \otimes \cdots \otimes x_n \mapsto$

$f(x_1)\dots f(x_n)$. But each X_i is finitely generated and projective, so it is easy to check that η is an isomorphism. This completes the proof. □

In this lemma the hypothesis that the X_i are projective is essential. (For a counter-example in the non-projective case, consider modules of order 3 over $\mathbf{Z}/9\,\mathbf{Z}$.)

As examples of this tensor product operation, note that

$$\langle u\rangle \otimes \langle v\rangle \cong \langle u\,v\rangle$$

for any u and v in $R^{\bullet}$, and that

$$\langle 1\rangle \otimes X \cong X$$

for any X. If X and Y are free modules, note that

$$\operatorname{rk}(X\otimes Y)=\operatorname{rk}(X)\operatorname{rk}(Y).$$

The tensor product operation will play a very important role in subsequent sections. In particular, it is used to provide the product operation in the Witt ring (§ 7). Here is a related construction which will also play an important role.

(5.4) Change of rings. Let $f\colon R\to R'$ be a ring homomorphism. Then any inner product space X over R gives rise to an inner product space

$$f_{\#}(X)=R'\underset{R}{\otimes} X$$

over the ring R', the inner product on $f_{\#}(X)$ being defined by the formula

$$(\alpha\otimes x)\cdot(\beta\otimes y)=\alpha\,\beta f(x\cdot y).$$

As an example, if X is free over R with basis $e_1, \dots, e_n$ and inner product matrix $(e_i\cdot e_j)$, then $f_{\#}(X)$ is free over R' with basis $1\otimes e_1, \dots, 1\otimes e_n$ and inner product matrix $\big(f(e_i\cdot e_j)\big)$. The correspondence $X\mapsto f_{\#}(X)$ preserves orthogonal sums and tensor products.

To conclude this section, we briefly describe three related constructions which will not play any important role in our presentation.

(5.5) The rank of a projective module. A similar construction is used to define the *rank* of a finitely generated projective module P over R. (See for example [Swan, 1968, p. 136].) Let $\operatorname{Spec}(R)$ be the collection of prime ideals in R with the Zariski topology, and let $\mathbf{Z}^{\operatorname{Spec}(R)}$ be the group of continuous integer valued functions on $\operatorname{Spec}(R)$. Then by definition

$$\operatorname{rank}(P)\in\mathbf{Z}^{\operatorname{Spec}(R)}$$

is the function which assigns to each prime ideal $\mathfrak{p}$ the dimension of the vector space $F\underset{R}{\otimes} P$, where F is the quotient field of $R/\mathfrak{p}$. This coincides with our previous definition whenever P is free.

(5.6) Exterior powers. If X is a bilinear form module over R, then the exterior power $\wedge^k X$ over R possesses a unique bilinear form $\hat{\beta}$ satisfying the identity

$$\hat{\beta}(x_1 \wedge \ldots \wedge x_k, y_1 \wedge \ldots \wedge y_k) = \det(\beta(x_i, y_j)).$$

See [Bourbaki, **24**, p. 30]. If X is ε-symmetric, then $\wedge^k X$ is ε^k-symmetric. If X is an inner product space, then $\wedge^k X$ is also an inner product space.

(5.7) The determinant. Let X be a bilinear form space over R whose underlying projective module has rank n at every prime ideal. Then the bilinear form space $\wedge^n X$ of rank 1 is called the "determinant" of X. Compare [Knebusch, 1969/70]. This subsumes our previous concept of determinant, since if X is free with basis $e_1, \ldots, e_n$, then $\wedge^n X$ is free with basis $e = e_1 \wedge \ldots \wedge e_n$, and hence

$$\wedge^n X \cong \langle d_0 \rangle,$$

where the element

$$d_0 = \hat{\beta}(e, e) = \det(\beta(e_i, e_j))$$

is well defined up to multiplication by squares of units.

An illustration of this concept of "determinant" will be given in Chapter V, § 3.

§ 6. Split Inner Product Spaces

(6.1) Definition. A symmetric inner product space S over the ring R is *split* if there exists a submodule $N \subset S$, such that N is a direct summand of S, and such that N is precisely equal to its orthogonal complement $N^\perp$.

This concept is due to Knebusch, who uses the term "metabolic" in place of our "split".

An equivalent form of the definition would be the following. The space S is split if it is the direct sum of two submodules M, N which are dually paired to R by the inner product,

$$M \xrightarrow{\cong} \operatorname{Hom}_R(N, R), \qquad N \xrightarrow{\cong} \operatorname{Hom}_R(M, R),$$

and such that N is self-orthogonal, $N \cdot N = 0$. [Still another formulation would be that S is split if it contains a self-orthogonal direct summand N whose rank, in the sense of § 5.5, is equal to $\frac{1}{2}\operatorname{rank}(S)$.]

Here are two examples. The *hyperbolic plane*, with inner product matrix $\begin{pmatrix} 0 & 1 \\ 1 & 0 \end{pmatrix}$, is clearly split. Furthermore, for any unit u, the orthogonal sum

$$\langle u \rangle \oplus \langle -u \rangle$$

is split. For if e_1, e_2 is an orthogonal basis with inner product matrix $\begin{pmatrix} u & 0 \\ 0 & -u \end{pmatrix}$, then the element e_1+e_2 spans the required direct summand N with $N=N^{\perp}$.

(6.2) Lemma. *Let S and S' be split inner product spaces, and let $X=(X,\beta)$ be an arbitrary inner product space. Then the orthogonal sum $S\oplus S'$ is split, and the tensor product $S\otimes X$ is split. Furthermore the orthogonal sum*

$$(X,\beta)\oplus(X,-\beta)$$

is split.

Proofs of the first two assertions are easily supplied. To prove the third, we use the split inner product space $\langle 1\rangle\oplus\langle -1\rangle$ to conclude that the tensor product

$$\begin{aligned}(\langle 1\rangle\oplus\langle -1\rangle)\otimes(X,\beta) &\cong (\langle 1\rangle\otimes(X,\beta))\oplus(\langle -1\rangle\otimes(X,\beta))\\ &\cong (X,\beta)\oplus(X,-\beta)\end{aligned}$$

is also split. □

In special cases we can give a more precise description of split spaces.

(6.3) Lemma. *Let R be a ring such that every finitely generated projective over R is free. Then an inner product space over R is split if and only if it possesses a basis so that the associated inner product matrix has the form $\begin{pmatrix} 0 & I \\ I & A \end{pmatrix}$. If we also assume that 2 is a unit, then every split inner product space has matrix $\begin{pmatrix} 0 & I \\ I & 0 \end{pmatrix}$ with respect to a suitable basis.*

Thus every split space is isomorphic to an orthogonal sum of hyperbolic planes, when all of the hypotheses of (6.3) are satisfied. This is the case when R is a field of characteristic $\neq 2$.

Proof. Given any direct summand $N\subset X$, choose a basis $e_1,\ldots,e_n$ for N, and extend to a basis $e_1,\ldots,e_k$ for X. Let $e_1^{\#},\ldots,e_k^{\#}$ be the dual basis. Then clearly $e_{n+1}^{\#},\ldots,e_k^{\#}$ form a basis for the orthogonal complement $N^{\perp}$.

Suppose that $N=N^{\perp}$. Then substituting $e_1,\ldots,e_n$ for $e_{n+1}^{\#},\ldots,e_k^{\#}$, we see that the elements

$$e_1,\ldots,e_n,\quad e_1^{\#},\ldots,e_n^{\#}$$

form a basis for X. (In particular the rank k of X must be equal to $2n$.) The inner product matrix of X with respect to this new basis takes the form $\begin{pmatrix} 0 & I \\ I & A \end{pmatrix}$ for some symmetric matrix A. The converse is clear.

Now suppose that 2 is a unit in R. Setting $B=-\frac{1}{2}A$, computation shows that

$$\begin{pmatrix} I & 0 \\ B & I \end{pmatrix} \begin{pmatrix} 0 & I \\ I & A \end{pmatrix} \begin{pmatrix} I & 0 \\ B & I \end{pmatrix}^t = \begin{pmatrix} 0 & I \\ I & 0 \end{pmatrix}.$$

This completes the proof. □

§ 7. The Witt Ring

Following Knebusch, we bring some order into the collection of all inner product spaces over R by introducing an equivalence relation. (See also [Fröhlich-McEvett].)

(7.1) Definition. Two symmetric inner product spaces X and X' over R belong to the same *Witt class*, written $X \sim X'$, if there exist split inner product spaces S and S' so that $X \oplus S$ is isomorphic to $X' \oplus S'$.

Evidently this is an equivalence relation. Furthermore:

(7.2) Lemma. *If $X \sim X'$ and $Y \sim Y'$, then $X \oplus Y \sim X' \oplus Y'$ and $X \otimes Y \sim X' \otimes Y'$.*

Proof. These statements follow easily from (6.2). □

Now recalling that $(X, \beta) \oplus (X, -\beta) \sim 0$ and $\langle 1 \rangle \otimes X \cong X$, we evidently obtain the following.

(7.3) Theorem. *The collection $W(R)$ of all Witt classes of symmetric inner product spaces over R forms a commutative ring with* 1, *using the orthogonal sum as addition operation and the tensor product as multiplication operation.*

Following Knebusch, $W(R)$ is called the *Witt ring* of R. Using (5.4) we see that any ring homomorphism $R \to R'$ induces a ring homomorphism $W(R) \to W(R')$. We will investigate the structure of this ring $W(R)$ in subsequent chapters.

Note the following.

(7.4) Lemma. *If R is a local ring in which* 2 *is a unit, then two symmetric inner product spaces over R are isomorphic if and only if they belong to the same Witt class and have the same rank.*

Proof. This follows easily from (4.4) and (6.3). □

Chapter II. Symmetric Inner Product Spaces over Z

This chapter will discuss the classification problem for inner product spaces over the ring **Z** of rational integers. All inner products are to be symmetric. Our presentation is based on the classical theorem of Minkowski concerning lattice points in a convex symmetric subset of $\mathbf{R}^n$. This theorem is first used to classify inner product spaces of rank $\leqq 4$ over **Z**. Making use of the Hasse-Minkowski theorem (which we do not prove), it is shown that an indefinite inner product space over **Z** is completely determined by its rank, type, and signature; where the *type* is defined to be either *I* or *II* according as the space does or does not contain a vector of odd norm. It follows that the Witt ring $W(\mathbf{Z})$ is isomorphic to **Z**.

The classification problem for positive definite inner product spaces, on the other hand, is extraordinarily difficult. After a discussion of this problem, and the related problem of close packing of balls in euclidean space, we present the classical characterization of sums of two or four squares in **Z**. The chapter concludes with an outline of Siegel's work on positive definite bilinear forms over **Z**.

§ 1. Minkowski's Convex Body Theorem

Let $\mathbf{R}^n$ be the cartesian space consisting of all n-tuples $x=(x_1, \ldots, x_n)$ of real numbers, and provided with the standard Lebesgue measure $dx_1 \ldots dx_n$.

(1.1) Definition. A *lattice* in $\mathbf{R}^n$ is an additive subgroup $L \subset \mathbf{R}^n$ which is additively generated by some basis $b_1, \ldots, b_n$ for the real vector space $\mathbf{R}^n$.

Choosing some basis $b_1, \ldots, b_n$ for L, we can form the fundamental domain P consisting of all $\xi_1 b_1 + \cdots + \xi_n b_n$ with $0 \leqq \xi_i < 1$. Clearly every point of $\mathbf{R}^n$ is congruent modulo L to one and only one point of P. The volume (or Lebesgue measure)

$$\operatorname{vol}(P) = \int_P dx_1 \ldots dx_n$$

can be identified with the volume of the quotient torus $\mathbf{R}^n/L$. This volume is of course equal to the absolute value of the determinant of the matrix whose rows are $b_1, \ldots, b_n$. (See for example [Birkhof-MacLane].) We write this briefly as

$$\operatorname{vol}(\mathbf{R}^n/L) = |\det(b_1, \ldots, b_n)|.$$

A lattice is called *unimodular* if $\mathbf{R}^n/L$ has volume 1.

(1.2) Examples. Clearly $\mathbf{Z}^n \subset \mathbf{R}^n$ is a lattice with $\operatorname{vol}(\mathbf{R}^n/\mathbf{Z}^n) = 1$. If L and L' are lattices with $L \supset L'$, then clearly the index $|L/L'|$ is finite and

$$\operatorname{vol}(\mathbf{R}^n/L') = \operatorname{vol}(\mathbf{R}^n/L)\,|L/L'|.$$

If we think of $\mathbf{R}^n$ as a euclidean inner product space, with inner product $x \cdot y = \sum x_i y_i$, then the volume $|\det(b_1, \ldots, b_n)|$ can also be written as $\sqrt{\det(b_i \cdot b_j)} = \sqrt{\det L}$. Note that any inner product space X over $\mathbf{Z}$ embeds canonically as a lattice in the real inner product space $\mathbf{R} \otimes X$. If X is positive definite (that is if $x \cdot x > 0$ for $x \neq 0$), then $\mathbf{R} \otimes X$ must be isomorphic to the euclidean space $\mathbf{R}^n$, and X corresponds to a unimodular lattice in $\mathbf{R}^n$.

Recall that a subset $K \subset \mathbf{R}^n$ is *convex* if $x, x' \in K$ implies that $\lambda x + (1-\lambda) x' \in K$ for all real numbers λ in the interval $0 \leqq \lambda \leqq 1$. A subset K of $\mathbf{R}^n$ is *symmetric about* 0, if $x \in K$ implies $-x \in K$.

(1.3) Minkowski's theorem. *Let K be a convex subset of $\mathbf{R}^n$ which is symmetric about* 0. *If the volume (or Lebesgue measure) of K is greater than 2^n times the volume of a fundamental domain for L, then K contains a non-zero lattice point.*

Proof. The subset K' consisting of all $\frac{1}{2}x$ with $x \in K$ clearly satisfies

$$\operatorname{vol}(K') > \operatorname{vol}(\mathbf{R}^n/L).$$

Hence the canonical map $K' \to \mathbf{R}^n/L$, which is locally a volume preserving embedding, cannot be one-to-one. There must exist two distinct points say $\frac{1}{2}x$ and $\frac{1}{2}y$ in K' with the same image in $\mathbf{R}^n/L$; so that

$$0 \neq \tfrac{1}{2}x - \tfrac{1}{2}y \in L.$$

But $\frac{1}{2}(x-y)$ is the midpoint of two points x and $-y$ of K, and hence itself belongs to K. This completes the proof. □

Example. Let $D(r)$ denote the closed euclidean disk of radius r in $\mathbf{R}^n$. Then the volume of $D(r)$ is equal to $\omega_n r^n$ where the numbers

$$\omega_1 = 2, \quad \omega_2 = \pi, \quad \omega_3 = \frac{4}{3}\pi, \quad \omega_4 = \frac{\pi^2}{2}, \ldots$$

can be computed inductively by the formula

$$\omega_n = \int_0^{2\pi} \int_0^1 \omega_{n-2} (\sqrt{1-r^2})^{n-2}\, r\, dr\, d\theta = 2\pi\, \omega_{n-2}/n.$$

It follows that $\omega_n = \pi^{n/2}/(n/2)!$ for even values of n. (In terms of the gamma function, we have $\omega_n = \pi^{n/2}/\Gamma(1+\frac{1}{2}n)$ for all values of n.) Applying Minkowski's theorem to $D(r)$, we obtain the following.

(1.4) Corollary. *If* $r^n \geqq (2^n/\omega_n)\operatorname{vol}(\mathbf{R}^n/L)$, *then the disk* $D(r)$ *contains a non-zero lattice point.*

Proof. If r^n is strictly larger than $(2^n/\omega_n)\operatorname{vol}(\mathbf{R}^n/L)$, this follows immediately. If equality holds, we simply note that $D(r+\varepsilon)$ contains a non-zero lattice point for every $\varepsilon > 0$. But the compact set $D(r+\varepsilon)$ can contain only finitely many non-zero lattice points, so the closest one to the origin must lie in $D(r)$. □

Setting r_0 precisely equal to the n-th root of $(2^n/\omega_n)\operatorname{vol}(\mathbf{R}^n/L)$, this yields the following.

(1.5) Corollary. *Every lattice* $L \subset \mathbf{R}^n$ *contains a point* $(x_1, \ldots, x_n)$ *with*

$$0 < x_1^2 + \cdots + x_n^2 \leqq r_0^2 = 4(\omega_n^{-1} \operatorname{vol}(\mathbf{R}^n/L))^{2/n}.$$

In particular every unimodular lattice in $\mathbf{R}^n$ *contains a point with*

$$0 < x_1^2 + \cdots + x_n^2 \leqq 4/(\omega_n)^{2/n}.$$

By Stirling's formula, this upper bound $4/(\omega_n)^{2/n}$ is asymptotic to $2n/\pi e$ as $n \to \infty$. (Compare § 7.) Here is a table listing values of $4/(\omega_n)^{2/n}$ rounded to the nearest hundredth.

n	1	2	3	4	5	6	7	8	9	10	11
$4/(\omega_n)^{2/n}$	1	1.27	1.54	1.80	2.06	2.31	2.57	2.82	3.07	3.32	3.56

Note in particular that $4/(\omega_n)^{2/n} < 2$ whenever $n \leqq 4$.

Remark. More generally, for any positive integer n one has the estimate

$$4/(\omega_n)^{2/n} < 1 + \tfrac{1}{4} n.$$

For $n < 12$ this can be verified by inspecting the table, and for $n \geqq 12$ it can be proved as follows. Setting $A_n = (1+\frac{1}{4}n)^{n/2}\, \omega_n/2^n$ and setting $n = 2u - 2$, we have

$$A_n/A_{n-2} = \frac{\pi}{8} \frac{u^2}{u^2-1} \left(1 + \frac{1}{u}\right)^u$$

where the expression $\frac{\pi}{8}\left(1+\frac{1}{u}\right)^u$ is monotone increasing for $u \geqq 2$ and greater than 1 for $u=7$. It follows inductively that $A_n > A_{n-2} > 1$ for $n \geqq 12$. $\square$

We will describe some non-trivial examples in § 6; and describe a sharper upper bound for the number

$$\min_{\substack{x \in L \\ x \neq 0}} (x_1^2 + \cdots + x_n^2)$$

in § 7.

Here is an important qualitative result which follows from (1.5).

(1.6) Lemma (Eisenstein, Hermite). *For each integer n there exist only finitely many distinct isomorphism classes of positive definite inner product spaces of rank n over* **Z**.

In fact, given positive integers n and d, there exist only finitely many distinct positive definite bilinear form spaces over **Z** with rank n and determinant d. (Compare § 7.4.) The proof, by induction on n, can be outlined as follows. Clearly any such space can be embedded isometrically as a lattice $L \subset \mathbf{R}^n$, where $\mathbf{R}^n$ has the euclidean inner product. The volume of a fundamental domain for L will be $\sqrt{d}$. Using (1.5), there is a constant $c(n, d) = 4\sqrt[n]{d/\omega_n^2}$ so that every such lattice L contains a vector x with

$$0 < x \cdot x \leqq c(n, d).$$

Let L_0 be the sublattice, of index at most $x \cdot x$, consisting of all $y \in L$ with

$$x \cdot y \equiv 0 \quad \mod x \cdot x.$$

Then L_0 decomposes as the orthogonal sum of the subgroup spanned by x, and its orthogonal complement. Using the induction hypothesis, there exist only finitely many possibilities for L_0, up to isomorphism. Hence there exist only finitely many possibilities for L. $\square$

A similar argument can be given for indefinite bilinear forms.

§ 2. Inner Product Spaces of Rank $\leqq 4$ over Z

(2.1) Lemma. *If L is an inner product space of rank n over* **Z**, *then L contains a vector* $x \neq 0$ *with*

$$|x \cdot x| \leqq 4/(\omega_n)^{2/n}.$$

In particular, if $n \leqq 4$ *then L contains a vector* $x \neq 0$ *with* $|x \cdot x| < 2$.

Proof. Note that L embeds isometrically as a lattice in the real inner product space $\mathbf{R}\otimes L$. If L is positive definite, then we can identify $\mathbf{R}\otimes L$ with euclidean n-space, and apply (1.5) directly. In any case, it is possible to choose an orthogonal basis $e_1, \ldots, e_n$ for $\mathbf{R}\otimes L$ so that $e_i \cdot e_i = \pm 1$. Using this basis to identify $\mathbf{R}\otimes L$ with the cartesian space $\mathbf{R}^n$, and to introduce the volume element $dx_1 \ldots dx_n$, note that the volume of $(\mathbf{R}\otimes L)/L$ is equal to 1. For if $b_1, \ldots, b_n$ is a basis for L, then the matrix equation

$$(b_i \cdot b_j) = \begin{pmatrix} b_1 \\ \vdots \\ b_n \end{pmatrix} \begin{pmatrix} \pm 1 & & \\ & \ddots & \\ & & \pm 1 \end{pmatrix} (b_1^t \ldots b_n^t)$$

implies that $\det(b_i \cdot b_j) = \pm 1$ is equal to $\pm \det(b_1, \ldots, b_n)^2$.

Therefore, by (1.5), there exists a lattice element $x = x_1 e_1 + \cdots + x_n e_n \neq 0$ with

$$x_1^2 + \cdots + x_n^2 \leqq 4/(\omega_n)^{2/n}.$$

Evidently it follows that

$$|x \cdot x| = |\pm x_1^2 \pm \cdots \pm x_n^2| \leqq 4/(\omega_n)^{2/n},$$

as required. Since it is easy to check that $4/(\omega_n)^{2/n} < 2$ for $n \leqq 4$, this completes the proof. □

(2.2) Theorem. *Every inner product space of rank $\leqq 4$ over $\mathbf{Z}$ either possesses an orthogonal basis, and hence is isomorphic to a sum of copies of $\langle 1\rangle$ and $\langle -1\rangle$, or is "hyperbolic", with inner product matrix of the form* $\begin{pmatrix} 0 & I \\ I & 0 \end{pmatrix}$.

Proof. We argue by induction. The statement is certainly true in the rank 1 case. Suppose then that the rank is $n > 1$. If we can find a vector x in X with $x \cdot x = \pm 1$, then evidently

$$X \cong \langle \pm 1\rangle \oplus X'.$$

If X' has an orthogonal basis, we are finished; while if X' is hyperbolic, spanned by y and z with $y \cdot y = z \cdot z = 0$, $y \cdot z = 1$, then the vectors $x + y$, $x \mp z$, and $x + y \mp z$ form the required orthogonal basis.

Suppose on the other hand that we can find a vector $x_1 \neq 0$ with $x_1 \cdot x_1 = 0$. Without loss of generality, we may assume that x_1 is indivisible, so that x_1 forms part of a basis $x_1, \ldots, x_n$ for X. Let $y_1, \ldots, y_n$ be the dual basis. Then the subspace spanned by x_1 and y_1 has an inner product matrix of the form $\begin{bmatrix} 0 & 1 \\ 1 & a \end{bmatrix}$. There are two possibilities.

Case 1. If the entry $a=y_1 \cdot y_1$ is even, then the subspace spanned by x_1 and y_1 is hyperbolic, with a basis x_1 and $y_1-\frac{1}{2}a\,x_1$ consisting of self-orthogonal vectors. The inner product matrix with respect to this new basis is $\begin{bmatrix}0 & 1\\ 1 & 0\end{bmatrix}$. Thus X is isomorphic to an orthogonal sum

$$\left\langle\begin{bmatrix}0 & 1\\ 1 & 0\end{bmatrix}\right\rangle \oplus X';$$

and applying the induction hypothesis to X' the proof is easily completed.

Case 2. If $a=2k+1$ is odd, then the two vectors

$$x'=y_1-k\,x_1, \qquad y'=y_1-(k+1)\,x_1$$

are mutually orthogonal, with inner product matrix $\begin{bmatrix}1 & 0\\ 0 & -1\end{bmatrix}$. So again we can split off an orthogonal summand and proceed by induction.

Since, by (2.1), there always exists a vector $x\in X$ with either $x\cdot x=\pm 1$ or $x\cdot x=0$, this completes the proof. □

(2.3) Remark. The statement of (2.2) is actually true for ranks 5, 6, and 7 also. It is false for ranks $\geqq 8$, as we will see in § 6.

§ 3. The Hasse-Minkowski Theorem and Meyer's Theorem

Next we will invoke a basic theorem of algebraic number theory, which we will not prove. Given an inner product space X over the rational numbers **Q**, we want to find a vector $x\neq 0$ with $x\cdot x=0$. Choosing an orthogonal basis, so that

$$X\cong\langle a_1\rangle\oplus\cdots\oplus\langle a_n\rangle$$

we see that this is equivalent to finding a non-trivial solution to the equation

$$a_1\,\xi_1^2+\cdots+a_n\,\xi_n^2=0.$$

(3.1) Theorem (Hasse-Minkowski). *The equation $a_1\,\xi_1^2+\cdots+a_n\,\xi_n^2=0$, with non-zero rational coefficients, has a non-trivial rational solution if and only if* (1) *it has a non-trivial real solution and* (2) *for every prime number p it has a non-trivial solution in the field $\mathbf{Q}_p$ of p-adic numbers.*

Proofs are given for example in [Borevich-Shafarevich], [O'Meara] and [Serre]. An outline of a proof is given in Appendix 3.

(3.2) Corollary (Meyer's theorem). *An indefinite inner product space of rank $n\geqq 5$ over* **Q** *always possesses a vector $x\neq 0$ with $x\cdot x$ equal to* 0.

(The word "indefinite" means that the norm $x \cdot x$ takes on both positive and negative values. The restriction $n \geqq 5$ is essential. For example the equation $\xi_1^2 + \xi_2^2 + \xi_3^2 - 7\xi_4^2$ has no non-trivial rational solution as can easily be verified by clearing denominators and then reducing modulo 8.)

To prove the corollary from the theorem, we need only show that the equation

$$a_1 \xi_1^2 + \cdots + a_5 \xi_5^2 = 0$$

has a non-trivial p-adic solution. We use the following lemmas which are also referred to later.

(3.3) Lemma. *If F is a finite field, then for any choice a, b, c of non-zero elements of F the equation*

$$a\xi^2 + b\eta^2 = c$$

has a solution in F.

The proof will be based on the "shoe box principle." Let q be the number of elements in F. We may assume that q is odd, since otherwise the equation $a\xi^2 = c$ already has a solution.

As ξ varies over F, note that the expression $a\xi^2$ takes on $\frac{1}{2}(q+1)$ distinct values. Similarly, as η varies over F the expression $c - b\eta^2$ takes on $\frac{1}{2}(q+1)$ distinct values. Since

$$\tfrac{1}{2}(q+1) + \tfrac{1}{2}(q+1) > q,$$

it follows that at least one of these values $a\xi^2$ must be equal to one of the values $c - b\eta^2$. Hence $a\xi^2 + b\eta^2 = c$ as required. □

(3.4) Lemma. *Let p be an odd prime, and let u, v, w be units in the ring $\mathbf{Z}_p$ of p-adic integers. Then the equation $u\xi^2 + v\eta^2 + w\zeta^2 = 0$ has a non-trivial solution with $\xi, \eta, \zeta \in \mathbf{Z}_p$.*

Proof. Since p is odd, any number congruent to 1 modulo p is a p-adic square. In fact let U_k denote the subgroup of $\mathbf{Z}_p^\bullet$ consisting of all units which are congruent to 1 modulo $p^k \mathbf{Z}_p$. Then the quotient U_1/U_k is a finite abelian group of odd order p^{k-1}; so clearly every element of U_1 has a unique square root modulo U_k. Taking the limit, as $k \to \infty$, of this square root modulo U_k, we obtain an actual square root in U_1.

Now given u, v, w and using (3.3) we can find p-adic integers ξ and η with

$$u\xi^2 + v\eta^2 \equiv -w \mod p,$$

or in other words

$$-(u\xi^2 + v\eta^2)/w \equiv 1 \mod p.$$

Setting

$$\zeta = \sqrt{-(u\xi^2 + v\eta^2)/w},$$

it follows that $u\xi^2 + v\eta^2 + w\zeta^2 = 0$ as required. □

Proof of (3.2). Given $a_1, \ldots, a_5 \in \mathbf{Q}_p$, we may assume that each a_i is either a p-adic unit or p times a unit. If at least three of the a_i are units, we are finished by (3.4). But otherwise at least three of the a_i are equal to p times a unit, and again it follows from (3.4) that there is a p-adic solution.

The proof for $p=2$ is similar, but one must work modulo 8, since a number must be congruent to 1 modulo 8 in order to guarantee that it is a 2-adic square. We may assume, after permuting the coefficients and multiplying by a constant, that $a_1=1$, that a_2 and a_3 are 2-adic units, and that a_4 and a_5 are divisible by at most 2. It is then not difficult to check that the congruence

$$a_2\,\xi_2^2+a_3\,\xi_3^2+a_4\,\xi_4^2+a_5\,\xi_5^2\equiv -1 \pmod 8$$

has a solution, and the argument proceeds as above. ∎

§ 4. Indefinite Spaces over Z

(4.1) Lemma. *Every indefinite inner product space over* **Z** *possesses a vector* $x\neq 0$ *with* $x\cdot x=0$.

Proof. If the rank is $\geqq 5$, this follows from Meyer's theorem (3.2), while if the rank is $\leqq 4$ it follows from (2.2). ∎

Remark. This argument depends of course on the Hasse-Minkowski theorem, which we have not proved. A different and self-contained proof of (4.1) will be given in Chapter IV.

(4.2) Definition. An inner product space over **Z** is of *type* I if it contains a vector x with $x\cdot x$ odd, and of *type* II if there is no such vector.

Evidently X is of type II if and only if the quadratic function $q(x)=\frac{1}{2}\,x\cdot x$ takes values in **Z**. Thus the inner product spaces of type II are precisely those which arise from quadratic inner product spaces over **Z**. (Compare Appendix 1.)

(4.3) Theorem. *Every indefinite inner product space of type* I *over* **Z** *possesses an orthogonal basis, and hence is isomorphic to an orthogonal sum of copies of* $\langle 1\rangle$ *and* $\langle -1\rangle$.

(It follows that such a space is uniquely determined by its rank and signature. Compare the discussion below.)

Proof by induction. The statement is already known for small values of the rank. Choose a vector $x_1\neq 0$ with $x_1\cdot x_1=0$. Without loss of generality we may assume that x_1 is indivisible, and hence forms part of a basis $x_1, \ldots, x_n$ for X. Let $y_1, \ldots, y_n$ be the dual basis. Then $x_1\cdot y_1=1$.

By hypothesis, there exists a vector y whose norm $y \cdot y$ is odd. Hence one of the basis vectors y_k must have odd norm. Choose a subspace $X_0 \subset X$ as follows. If $y_1 \cdot y_1$ is odd, let X_0 be spanned by x_1 and y_1. If $y_1 \cdot y_1$ is even, let X_0 be spanned by x_1 and $y_1 + y_k$ where k is chosen so that

$$(y_1 + y_k) \cdot (y_1 + y_k) \equiv y_k \cdot y_k \equiv 1 \quad \bmod 2.$$

Then X_0 has inner product matrix of the form

$$\begin{bmatrix} 0 & 1 \\ 1 & \text{odd} \end{bmatrix}.$$

As in the proof of (2.2), Case 2, we see that

$$X_0 \cong \langle 1 \rangle \oplus \langle -1 \rangle.$$

Clearly X splits as an orthogonal direct sum

$$X \cong X_0 \oplus X_0^\perp \cong \langle 1 \rangle \oplus \langle -1 \rangle \oplus X_0^\perp.$$

Choose a sign ± 1 so that $\langle \pm 1 \rangle \oplus X_0^\perp$ is indefinite. Then $\langle \pm 1 \rangle \oplus X_0^\perp$ possesses an orthogonal basis by the induction hypothesis, hence X does also and this proves the theorem. ▯

The *signature* of an inner product space X over **Z** is defined as follows. Choose an orthogonal basis $b_1, \ldots, b_n$ for the rational inner product space $\mathbf{Q} \otimes X$. Then the signature

$$\sigma(X) = \sigma(\mathbf{Q} \otimes X) \in \mathbf{Z}$$

is the difference between the number of basis elements b_i with $b_i \cdot b_i > 0$ and the number of basis elements b_j with $b_j \cdot b_j < 0$. The signature is an invariant of X by the Jacobi-Sylvester Inertia Theorem. (Compare Chapter III, § 2.5.) Note that $\sigma(X \oplus Y) = \sigma(X) + \sigma(Y)$.

The Witt ring $W(\mathbf{Z})$ can now be computed as follows.

(4.4) Corollary. *The ring $W(\mathbf{Z})$ is canonically isomorphic to* **Z**.

In fact, we will show that the correspondence $X \mapsto \sigma(X)$ gives rise to a ring isomorphism $W(\mathbf{Z}) \to \mathbf{Z}$. (Compare Chapter III, § 2.6; as well as Chapter IV, § 2.)

If S is a split inner product space over **Z**, then the signature $\sigma(S)$ *is* zero. For the rational inner product space $\mathbf{Q} \otimes S$ is isomorphic to a sum of copies of $\langle 1 \rangle \oplus \langle -1 \rangle$, using Chapter I, § 6.3; hence

$$\sigma(S) = \sigma(\mathbf{Q} \otimes S) = 0.$$

It follows easily that the correspondence $X \mapsto \sigma(X)$ gives rise to an additive homomorphism from $W(\mathbf{Z})$ to **Z**.

This homomorphism is surjective since $\sigma\langle 1\rangle=1$. It has kernel zero, since if $\sigma(X)=0$ then the sum $X\oplus\langle 1\rangle\oplus\langle -1\rangle$ is isomorphic to a sum of copies of $\langle 1\rangle\oplus\langle -1\rangle$ by (4.3), and therefore

$$X\sim X\oplus\langle 1\rangle\oplus\langle -1\rangle\sim 0.$$

Thus two inner product spaces over $\mathbf{Z}$ *have the same signature if and only if they belong to the same Witt class.* Since the identity $\sigma\langle 1\rangle=1$ implies that the bijection $W(\mathbf{Z})\to\mathbf{Z}$ is a ring isomorphism, this completes the proof. ∎

§ 5. Spaces of Type II

We will first prove the following.

(5.1) Theorem. *The signature of an inner product space of type* II *is necessarily divisible by* 8.

An example in which the signature is precisely 8 will be described in §6.

To prove (5.1), we first consider an arbitrary inner product space X over $\mathbf{Z}$. An element $u\in X$ will be called *characteristic* if

$$u\cdot x\equiv x\cdot x \pmod 2$$

for every x.

(5.2) Lemma (van der Blij). *Every inner product space* X *over* $\mathbf{Z}$ *possesses a characteristic element. Furthermore, if* $u\in X$ *is characteristic, then*

$$u\cdot u\equiv\sigma(X) \pmod 8.$$

Proof. Form the induced inner product space $X\otimes\mathbf{F}_2=X/2X$ over the field with 2 elements. If $\bar{x}$ denotes the image of x in $X/2X$, then the inner product $x\cdot y$ on X clearly gives rise to an $\mathbf{F}_2$-valued inner product

$$\bar{x}\cdot\bar{y}=(\text{residue class of } x\cdot y \bmod 2)$$

on $X/2X$.

The function $\bar{x}\mapsto\bar{x}\cdot\bar{x}$ from $X/2X$ to $\mathbf{F}_2$ is $\mathbf{F}_2$-linear. Hence there is one and only one element $\bar{u}\in X/2X$ which satisfies the equation

$$\bar{u}\cdot\bar{x}=\bar{x}\cdot\bar{x}$$

for all $\bar{x}$. Choosing any preimage $u\in X$, we obtain the required characteristic element.

Next note that the residue class of $u\cdot u$ modulo 8 is an invariant of X. For if u' is another characteristic element, then $u'=u+2x$, hence

$$u'\cdot u'=u\cdot u+4(u\cdot x+x\cdot x)\equiv u\cdot u \pmod 8.$$

Clearly this invariant $u \cdot u$ is additive with respect to direct sums. For the inner product space $\langle 1 \rangle$ note that $u \cdot u \equiv 1 \bmod 8$, while for $\langle -1 \rangle$ note that $u \cdot u \equiv -1 \bmod 8$. So for the orthogonal sum of p copies of $\langle 1 \rangle$ and q copies of $\langle -1 \rangle$ we see that $u \cdot u$ is congruent to the signature $\sigma = p - q \bmod 8$.

For any inner product space X, the orthogonal sum $X \oplus \langle 1 \rangle \oplus \langle -1 \rangle$ is indefinite of type I, and hence is isomorphic to a sum of copies of $\langle 1 \rangle$ and $\langle -1 \rangle$. Since the invariant $u \cdot u \bmod 8$ associated with $X \oplus \langle 1 \rangle \oplus \langle -1 \rangle$ is the residue class of $\sigma(X \oplus \langle 1 \rangle \oplus \langle -1 \rangle) = \sigma(X) \bmod 8$, and since the invariant associated with $\langle 1 \rangle \oplus \langle -1 \rangle$ is zero, this proves the lemma. ☐

Proof of (5.1). If X is of type II, then we can choose $u = 0$. Hence

$$\sigma(X) \equiv 0 \cdot 0 \qquad (\bmod 8),$$

which completes the proof. ☐

Remark 1. The invariant $u \cdot u$ modulo 8 is also defined for an inner product space over the ring $\mathbf{Z}_2$ of 2-adic integers. It gives rise to a homomorphism from the Witt ring $W(\mathbf{Z}_2)$ to $\mathbf{Z}/8\,\mathbf{Z}$. Evidently the diagram

$$\begin{array}{ccc} W(\mathbf{Z}) & \longrightarrow & W(\mathbf{Z}_2) \\ \downarrow \sigma & & \downarrow u \cdot u \\ \mathbf{Z} & \longrightarrow & \mathbf{Z}/8\,\mathbf{Z} \end{array}$$

is a commutative diagram of ring homomorphisms.

Remark 2. A different and quite intriguing formula for the signature mod 8 has recently been given by J. Milgram, in connection with a problem in topology. Let V be any inner product space of signature σ over the rational numbers, and let $L \subset V$ be any lattice which is small enough so that $l \cdot l \in 2\mathbf{Z}$ for every $l \in L$. If $L^{\#} \subset V$ denotes the "dual lattice", consisting of all $v \in V$ with $v \cdot L \subset \mathbf{Z}$, and if $v_1, \ldots, v_d$ is a complete set of coset representatives for $L^{\#}$ modulo L, then Milgram proves that

$$\exp(2\pi i \sigma/8) = \sum_{j=1}^{d} \exp(\pi i v_j \cdot v_j)/\sqrt{d}.$$

See Appendix 4 for further details. As an example, if L is unimodular then $L = L^{\#}$ and $d = 1$, so the right hand side of this equation is equal to 1, and it follows once more that $\sigma \equiv 0 \bmod 8$. This is very close to the original proof of this congruence in [van der Blij].

To conclude this section, we prove the following.

(5.3) Theorem. *If two indefinite inner product spaces over* $\mathbf{Z}$ *have the same rank, type, and signature, then they are isomorphic.*

For type I this was proved in § 4, so we need only consider type II.

Proof (following [Serre]). First consider the following construction. Given an inner product space X of type I, we can form the subset X_0 consisting of all x with $x \cdot x \equiv 0 \pmod 2$. Evidently this is a sublattice of index 2 in X. We would like to construct an inner product space of type II which also contains X_0 as a sublattice of index 2. If such a space exists, then evidently it must be contained in the "dual lattice"

$$X_0^{\#} \subset \mathbf{Q} \otimes X,$$

consisting of all vectors $x^{\#}$ in the vector space $\mathbf{Q} \otimes X$ which satisfy the condition

$$x^{\#} \cdot x \in \mathbf{Z}$$

for all $x \in X_0$. (Compare Chapter IV, § 3.) Since $X_0^{\#}$ contains X_0 as a sublattice of index 4, there are at most three different lattices which properly contain X_0 and are properly contained in $X_0^{\#}$. One of these is X itself. We will be interested in the other two.

As an example, let us apply this construction to the inner product space $W = \langle 1 \rangle \oplus \langle -1 \rangle$, with orthogonal basis e_1, e_2. Inspection shows that W_0 has basis $e_1 + e_2, e_1 - e_2$, and that $W_0^{\#}$ has a dual basis $\frac{1}{2}(e_1 - e_2)$, $\frac{1}{2}(e_1 + e_2)$. In this case there are three lattices lying between W_0 and $W_0^{\#}$. One of these is W, and the other two are evidently both isomorphic to the hyperbolic plane H.

Now consider the sum $Y \oplus W = Y \oplus \langle 1 \rangle \oplus \langle -1 \rangle$ where Y is an arbitrary inner product space of type II. Then the subspace $(Y \oplus W)_0$ of vectors of even norm is equal to $Y \oplus W_0$, hence $(Y \oplus W)_0^{\#} = Y \oplus W_0^{\#}$. Clearly there are three lattices lying between $(Y \oplus W)_0$ and $(Y \oplus W)_0^{\#}$. One is $Y \oplus W$, and the other two are both isomorphic to $Y \oplus H$.

Let Y' be another inner product space of type II having the same rank and signature as Y. Then

$$Y \oplus W \cong Y' \oplus W$$

by Theorem (4.3). Applying the construction above to both sides, it follows that $Y \oplus H \cong Y' \oplus H$.

But a straightforward argument shows that any indefinite inner product space of type II is isomorphic to the orthogonal sum $Y \oplus H$ for some Y. (Compare Sections 2 and 4.) Evidently this completes the proof. □

§ 6. The Classification Problem for Positive Definite Spaces

First let us describe some examples of positive definite inner product spaces over $\mathbf{Z}$. Let $\mathbf{R}^{4m}$ denote the euclidean space with orthonormal basis $e_1, \ldots, e_{4m}$.

(6.1) Lemma. *The vectors e_i+e_j and $\frac{1}{2}(e_1+\cdots+e_{4m})$ span a lattice $\Gamma_{4m}\subset\mathbf{R}^{4m}$ which is an inner product space over* $\mathbf{Z}$.

Proof. Let L_0 be the sublattice of index 2 spanned by the e_i+e_j. Evidently L_0 can also be considered as a sublattice of index 2 in the lattice spanned by $e_1,\ldots,e_{4m}$. Therefore a fundamental domain for L_0 has volume 2; hence a fundamental domain for Γ_{4m} has volume 1. Since inspection shows that the inner product of any two elements of Γ_{4m} is an integer, this completes the proof. ∎

[More explicitly, Γ_{4m} can be described as the set of all linear combinations $\xi_1 e_1+\cdots+\xi_{4m}e_{4m}$ with $2\xi_i\in\mathbf{Z}$, $\xi_1\equiv\xi_2\equiv\cdots\equiv\xi_{4m}$ mod $\mathbf{Z}$, and $\xi_1+\cdots+\xi_{4m}\equiv 0$ mod $2\mathbf{Z}$.]

(6.2) Lemma. *This inner product space Γ_{4m} has type I if m is odd, and type II if m is even.*

Proof. Since each vector e_i+e_j has norm 2, and the vector $\frac{1}{2}(e_1+\cdots+e_{4m})$ has norm m, the conclusion follows easily. ∎

Thus the lattices $\Gamma_8,\Gamma_{16},\Gamma_{24},\ldots$ provide examples of positive definite inner product spaces of type II with signatures 8, 16, 24, ...

We can get some insight into the collection of all positive definite inner product spaces over $\mathbf{Z}$ as follows.

(6.3) Definition. An inner product space X is *indecomposable* provided it cannot be expressed as the orthogonal sum of two non-trivial subspaces.

(6.4) Theorem (Eichler). *Every positive definite inner product space over $\mathbf{Z}$ splits uniquely as an orthogonal sum of indecomposable spaces.*

Proof. (Compare [Kneser, 1954].) Call a vector $x\neq 0$ of X *minimal* provided x cannot be expressed as a sum $y+z$ of two strictly shorter vectors of X (vectors with $y\cdot y<x\cdot x$ and $z\cdot z<x\cdot x$). Evidently the procedure of expressing a vector as a sum of shorter vectors must stop after a finite number of steps, hence X is spanned (or generated) by its collection of minimal vectors. If X splits as an orthogonal sum $X_1\oplus X_2$, note that every minimal vector of X must belong either to X_1 or to X_2.

We say that two minimal vectors x and x' are *equivalent* provided there exists a finite sequence $x=x_0,x_1,\ldots,x_k=x'$ of minimal vectors with $x_{i-1}\cdot x_i\neq 0$ for $1\leq i\leq k$. Then each equivalence class spans a subspace of X, and clearly X is the orthogonal direct sum of these subspaces. Since this splitting is uniquely defined, the proof is complete. ∎

Note that the above construction is easy to carry out in practice. Here is an example, where details are left to the reader.

(6.5) Proposition. *The inner product space Γ_{4m} is indecomposable for $m \geqq 2$. The minimal vectors of Γ_{4m} are precisely those vectors of the form $\pm e_i \pm e_j$ or $\frac{1}{2}(\pm e_1 \pm \cdots \pm e_{4m})$ in Γ_{4m}.*

Of course Γ_4 is not indecomposable, since by (2.2) every positive definite space of rank 4 over **Z** must be isomorphic to $\langle 1 \rangle \oplus \langle 1 \rangle \oplus \langle 1 \rangle \oplus \langle 1 \rangle$.

Remark 1. Kneser has determined all indecomposable spaces of rank $\leqq 16$. Besides the spaces $\langle 1 \rangle$, Γ_8, Γ_{12}, and Γ_{16} which we have described, he showed that there is one additional irreducible space (of type I) in each of the dimensions 14, 15, 16. However one can scarcely hope to push such an explicit computation much further, since the number of distinct spaces grows very rapidly with the rank. For example, considering only positive definite inner product spaces of type II, one obtains the following table. (Compare the discussion following (9.7), and the Concluding Remark in Appendix 5, as well as [Serre, p. 94].)

Dimension	8	16	24	32	40
Number of distinct spaces	1	2	24	$\geqq 10^7$	$\geqq 10^{51}$

Remark 2. The two non-isomorphic spaces $\Gamma_8 \oplus \Gamma_8$ and Γ_{16} in dimension 16 have the following remarkable property. For each positive integer k, the number of solutions to the equation $x \cdot x = 2k$ with $x \in \Gamma_8 \oplus \Gamma_8$ is precisely equal to the number of solutions with $x \in \Gamma_{16}$. (See Witt, Hamburg. Abh. **14** (1941), 323–337, or [Serre, p. 177].) In either case the number of solutions is equal to $480 \sum_{d|k} d^7$.

Remark 3. The lattice Γ_8, first discovered by Korkine and Zolotareff [2a] in 1873, also occurs in the study of the exceptional Lie group E_8. This lattice is more symmetrical than the higher Γ_{4m} in the following sense. The minimal vectors of Γ_8 all have the same length, and in fact the automorphism group of Γ_8 permutes the 240 minimal vectors transitively. The subgroup consisting of orientation preserving automorphisms, modulo its center, is a simple group of order $174\,182\,400 = 2^{12} \cdot 3^5 \cdot 5^2 \cdot 7$. (Compare [Bourbaki, v. **34**, pp. 228, 270].)

Remark 4. This lattice in dimension 8 has an extremely interesting analogue in dimension 24, discovered by John Leech in 1967. The Leech lattice L is a positive definite inner product space of type II with the property that $x \cdot x \geqq 4$ for every $x \neq 0$. (See Appendix 5.) The group of automorphisms of L permutes the 196 560 minimal vectors transitively; and this automorphism group modulo its center is a simple group of order $4\,157\,776\,806\,543\,360\,000 = 2^{21} \cdot 3^9 \cdot 5^4 \cdot 7^2 \cdot 11 \cdot 13 \cdot 23$. ([Conway].)

[2a] Its existence was proved nonconstructively by Smith in 1867.

It is not known whether any such symmetrical lattices exist in higher dimensions. It would be natural to look next in dimension 48, since the theory of modular forms [Serre, pp. 171–178] indicates that an inner product space of type II with $x \cdot x \geqq 6$ for $x \neq 0$ must have dimension at least 48. (It is intriguing to note that $8=3^2-1$, $24=5^2-1$, $48=7^2-1$.)

§ 7. The Packing of Equal Balls in $\mathbf{R}^n$

The unimodular lattice Γ_8 has the property that $x \cdot x \geqq 2$ for every $x \neq 0$ in Γ_8, while the Leech lattice has the property that $x \cdot x \geqq 4$ for every $x \neq 0$. The n-fold tensor product $\Gamma_8 \otimes \cdots \otimes \Gamma_8$ provides an example of a unimodular lattice with $x \cdot x \geqq 2^n$ for every $x \neq 0$. (Compare § 9.6.) These examples suggest the following question. *Fixing n, what is the largest possible value for the minimum non-zero norm*

$$\min_{x \in L - 0} (x \cdot x)$$

for a unimodular lattice L in $\mathbf{R}^n$?

An alternative and completely equivalent formulation is the following. *What is the largest possible value for the ratio*

$$\mu(L) = (\min_{L-0} x \cdot x) / \sqrt[n]{\det L},$$

where now L is allowed to vary over all lattices in $\mathbf{R}^n$? This version is equivalent, since any lattice of maximal rank in $\mathbf{R}^n$ can be transformed into a unimodular lattice by a similarity transformation $x \mapsto x / \sqrt[2n]{\det L}$ which does not affect the ratio $\mu(L)$.

The following observation is classical. (Compare [Watson, pp. 29–31].)

(7.1) Lemma. *For each dimension n there exists a lattice L_n which maximizes the ratio $\mu(L) = (\min_{x \in L-0} x \cdot x) / \sqrt[n]{\det L}$. Furthermore, choosing a basis for L_n, the associated inner product matrix (multiplied by a real constant if necessary) is a matrix of rational numbers.*

Hence the maximal ratio $\mu(L_n)$ is the n-th root of a rational number. The proof of (7.1) will be given at the end of this section.

For $n \leqq 5$ the maximum value of $\mu(L)$ was determined by Korkine and Zolotareff, and for $n \leqq 8$ by Blichfeldt. These maximum values, for $1 \leqq n \leqq 8$, are equal to 1, $\sqrt{4/3}$, $\sqrt[3]{2}$, $\sqrt[4]{4}$, $\sqrt[5]{8}$, $\sqrt[6]{64/3}$, $\sqrt[7]{64}$, and 2 respectively. Rounded to three decimal places, the values are as follows. (Compare p. 36, Fig. 3.)

n	$\mu(L_n)$
1	1
2	1.155
3	1.260
4	1.414
5	1.516
6	1.665
7	1.811
8	2

In the case $n=2$ the maximal value $\mu(L_2)=\sqrt{4/3}$ is attained by the lattice of Fig. 1, with inner product matrix $\begin{pmatrix} 2 & 1 \\ 1 & 2 \end{pmatrix}$. We will refer to this as the *regular hexagonal lattice*, since the associated "Voronoï polyhedron", consisting of all points in $\mathbf{R}^2$ which are at least as close to the origin as to any other lattice point, is a regular hexagon.

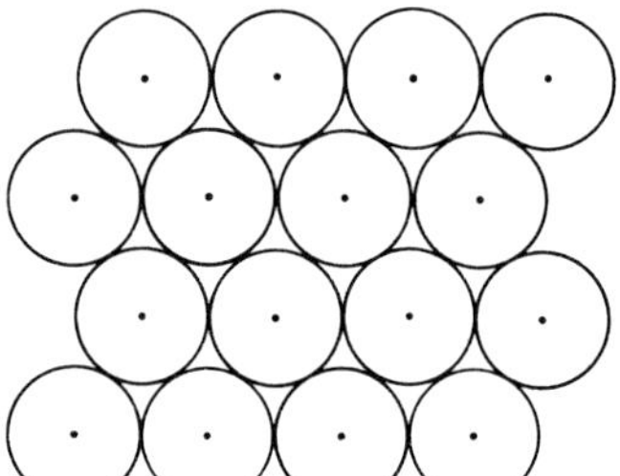

Fig. 1. The regular hexagonal lattice, and the associated packing of $\mathbf{R}^2$.

For $n=3$, 4, 5 the maximal value $\mu(L_n)$ is attained by a lattice L_n constructed as follows. Let $\mathbf{Z}^n$ be the lattice spanned by an orthonormal basis $e_1, \ldots, e_n$ for $\mathbf{R}^n$, and let L_n be the sublattice of index 2 spanned by the elements e_i+e_j. In the case $n=3$, this is known as the *face centered cubic* lattice, since it can be obtained from the "cubic" lattice with basis $2e_1$, $2e_2$, $2e_3$ by adjoining the center points of the faces of all of the cubes. (Compare Fig. 2, as well as [Hilbert and Cohn-Vossen].) The inner product matrix, with respect to the basis e_1+e_2, e_1+e_3, e_2+e_3 is

$$\begin{pmatrix} 2 & 1 & 1 \\ 1 & 2 & 1 \\ 1 & 1 & 2 \end{pmatrix}.$$

This lattice occurs in the real world as the configuration of atoms in a crystal of (for example) gold, silver, or aluminum. The Voronoï poly-

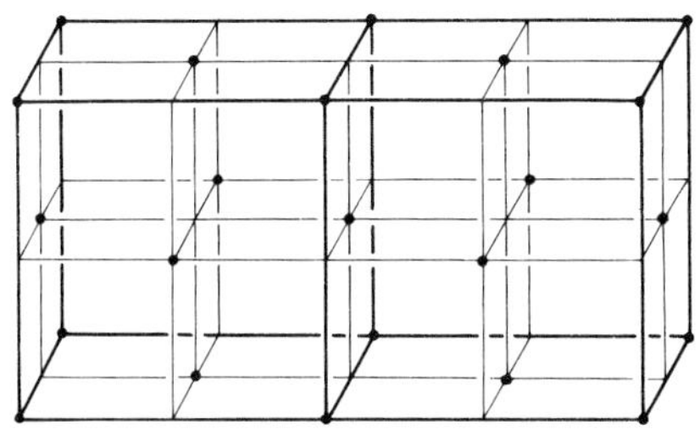

Fig. 2. The face centered cubic lattice in $\mathbf{R}^3$.

hedron associated with L_3 is a rhombic dodecahedron; that is, it is a solid bounded by twelve rhombuses.

For $n=6$, 7, 8 the function $\mu(L)$ is maximized by a lattice L_n constructed as follows. Let L_8 be the lattice Γ_8 of § 6, let L_7 be the orthogonal complement of a minimal vector in Γ_8, and let L_6 be the orthogonal complement of a regular hexagonal lattice in Γ_8. These three lattices can be identified with the lattices generated by the root systems of the exceptional Lie groups E_6, E_7, and E_8. (Similarly the lattices L_2, L_3, L_4, L_5 are generated by the root systems A_2, A_3, D_4, D_5. Compare pp. 138–139.)

For $n>8$, the precise value of $\mu(L_n)=\operatorname{Max}_{L\subset\mathbf{R}^n}\mu(L)$ is not known. However it can be computed up to a factor of 4 as follows. Recall that $\omega_n=\pi^{n/2}/\Gamma(1+n/2)$ denotes the volume of the unit disk in $\mathbf{R}^n$.

(7.2) Theorem of Minkowski. *The inequality*

$$\omega_n^{-2/n}<\mu(L_n)\leqq 4\,\omega_n^{-2/n}$$

is valid for every n.

In fact, the upper bound $\mu(L)\leqq 4\omega_n^{-2/n}$ is just a restatement of Corollary 1.5. Using Stirling's formula

$$\Gamma(1+x)\sim x^x e^{-x}\sqrt{2\pi x} \qquad \text{as } x\to\infty$$

we see that[3]

$$4\omega_n^{-2/n}\sim 2n/\pi e \qquad \text{as } n\to\infty.$$

A much sharper upper bound of the form

$$\mu(L)\leqq 4(\sigma_n/\omega_n)^{2/n}\sim n/\pi e$$

has been obtained by C.A. Rogers, and will be discussed later in this section. (Compare Fig. 3.)

The lower bound

$$\mu(L_n)>\omega_n^{-2/n}\sim n/2\pi e$$

[3] Or more precisely $\omega_n^{-2/n}=n/2\pi e+\log(\pi n)/2\pi e+o(1)$.

(or rather the slightly sharper lower bound $\mu(L_n) \geqq (2\zeta(n)/\omega_n)^{2/n}$) was proved by Minkowski in 1905. The name of E. Hlawka is often attached to this inequality since a generalization, stated by Minkowski, was first proved by Hlawka. Sharper inequalities of this form have been given by W. Schmidt, Rogers and others, but these all have the same asymptotic behavior as $n \to \infty$. Compare [Rogers, 1964]. A version involving self-dual lattices will be proved in § 9.5.

Proof of the inequality $\mu(L_n) \geqq (2/\omega_n)^{2/n} > \omega_n^{-2/n}$. The notation $x = (\xi_1, \ldots, \xi_n)$ will be used for a point in $\mathbf{R}^n$. Let $f(x) = f(\xi_1, \ldots, \xi_n) \geqq 0$ be a continuous real valued function with compact support. The integral $\int_{-\infty}^{\infty} \cdots \int_{-\infty}^{\infty} f(\xi_1, \ldots, \xi_n)\, d\xi_1 \ldots d\xi_n$ will be written briefly as $\int f(x)\, dx$. We will first prove the following. Assume that $n \geqq 2$.

(7.3) Lemma. *Given any real number* $\beta > \int f(x)\, dx$, *there exists a unimodular lattice* $L \subset \mathbf{R}^n$ *so that the sum* $\sum_{x \in L - 0} f(x)$ *is less than* β.

Proof. Let $e_1, \ldots, e_n$ be the standard orthonormal basis for $\mathbf{R}^n$. Let $\varepsilon > 0$ be a fixed small number, to be chosen later, and define $\lambda > 0$ by the equation $\varepsilon \lambda^{n-1} = 1$. Given real parameters $\tau_1, \ldots, \tau_{n-1}$ we consider the unimodular lattice $L = L(\tau_1, \ldots, \tau_{n-1})$ which is spanned by the basis

$$\lambda e_1, \ldots, \lambda e_{n-1}, \qquad \tau_1 \lambda e_1 + \cdots + \tau_{n-1} \lambda e_{n-1} + \varepsilon e_n.$$

Thus a typical element of this lattice $L(\tau_1, \ldots, \tau_{n-1})$ is the n-tuple

$$(\lambda(i_1 + j\tau_1), \ldots, \lambda(i_{n-1} + j\tau_{n-1}), j\varepsilon)$$

where $i_1, \ldots, i_{n-1}$ and j range independently over $\mathbf{Z}$. Clearly the lattice $L(\tau_1, \ldots, \tau_{n-1})$ remains unchanged if we add an integer to any one of the parameters τ_ν.

Fixing ε, for each value of the parameters $\tau_1, \ldots, \tau_{n-1}$ modulo 1 consider the sum

$$(1) \qquad \sum_{x \in L(\tau_1, \ldots, \tau_{n-1}) - 0} f(x) = \sum{}' f(\lambda(i_1 + j\tau_1), \ldots, \lambda(i_{n-1} + j\tau_{n-1}), j\varepsilon),$$

where the latter sum extends over all n-tuples $i_1, \ldots, i_{n-1}, j$ of integers, not all zero. Since f has compact support, we may choose ε so small that

$$f(\lambda i_1, \ldots, \lambda i_{n-1}, 0) = 0$$

for $(i_1, \ldots, i_{n-1}) \neq (0, \ldots, 0)$. If ε is so chosen, then the terms with $j = 0$ make no contribution, so we can rewrite the sum (1) as

$$(1') \qquad \sum_{j \neq 0} S_j(\tau_1, \ldots, \tau_{n-1})$$

where

$$S_j(\tau_1, \ldots, \tau_{n-1}) = \sum_{i_1, \ldots, i_{n-1}} f\big(\lambda(i_1 + j\tau_1), \ldots, \lambda(i_{n-1} + j\tau_{n-1}), j\varepsilon\big)$$

is zero for $|j|$ large. Next consider the average

$$(2) \qquad \begin{aligned} &\int_0^1 \cdots \int_0^1 \sum_{j \neq 0} S_j(\tau_1, \ldots, \tau_{n-1})\, d\tau_1 \ldots d\tau_{n-1} \\ &= \sum_{j \neq 0} \int_0^1 \cdots \int_0^1 S_j(\tau_1, \ldots, \tau_{n-1})\, d\tau_1 \ldots d\tau_{n-1} \end{aligned}$$

as $\tau_1, \ldots, \tau_{n-1}$ vary from 0 to 1. If $j > 0$, then the substitution $\eta_\nu = j\tau_\nu$ transforms the latter integral into

$$j^{1-n} \int_0^j \cdots \int_0^j \sum_{i_1 \ldots i_{n-1}} f\big(\lambda(i_1 + \eta_1), \ldots, \lambda(i_{n-1} + \eta_{n-1}), j\varepsilon\big)\, d\eta_1 \ldots d\eta_{n-1}.$$

But inspection shows that we are integrating precisely j^{n-1} times over each unit cube in $\mathbf{R}^{n-1}$, so that this expression is precisely equal to

$$\int_{-\infty}^{\infty} \cdots \int_{-\infty}^{\infty} f(\lambda\eta_1, \ldots, \lambda\eta_{n-1}, j\varepsilon)\, d\eta_1 \ldots d\eta_{n-1}.$$

A similar argument proves this formula when $j < 0$. Setting

$$g(\eta) = \int_{-\infty}^{\infty} \cdots \int_{-\infty}^{\infty} f(\xi_1, \ldots, \xi_{n-1}, \eta)\, d\xi_1 \ldots d\xi_{n-1},$$

this last expression is equal to

$$\lambda^{1-n} g(j\varepsilon) = \varepsilon\, g(j\varepsilon).$$

Therefore the average (2) is equal to

$$(3) \qquad \sum_{j \neq 0} \varepsilon\, g(j\varepsilon).$$

But the function $g(\eta)$ is continuous with compact support, so the sum (3) clearly converges, as $\varepsilon \to 0$, to the Riemann integral

$$\int_{-\infty}^{\infty} g(\eta)\, d\eta = \int f(x)\, dx < \beta.$$

Choosing ε so small that the Riemann sum (3) is less than β, it follows that the average (2) is also less than β. Therefore there must exist actual parameter values $\tau_1, \ldots, \tau_{n-1}$ so that the sum $(1) = (1')$ is less than β. This completes the proof of (7.3). □

To prove Theorem 7.2, we make a particular choice of the function f. Let $r < s$ be positive real numbers which are less than $(2/\omega_n)^{1/n}$. Choose f so that

$$f(x) = 1 \quad \text{for } x \cdot x \leqq r^2$$

$$f(x) = 0 \quad \text{for } x \cdot x \geqq s^2,$$

with $0 \leqq f(x) \leqq 1$ everywhere. Then evidently

$$\int f(x)\, dx < \omega_n s^n < 2.$$

Therefore, by (7.3), there exists a unimodular lattice $L \subset \mathbf{R}^n$ with

$$\sum_{x \in L-0} f(x) < 2. \tag{4}$$

This lattice L cannot contain any vector x_0 with

$$0 < x_0 \cdot x_0 \leqq r^2.$$

For if such a vector x_0 existed, then the summation (4) would include the terms $f(x_0) + f(-x_0) \geqq 2$, which is impossible. Therefore

$$\mu(L) = \operatorname*{Min}_{x \in L-0} x \cdot x > r^2.$$

Since r^2 can be any number less than $(2/\omega_n)^{2/n}$, this proves that

$$\mu(L_n) = \sup_{L \subset \mathbf{R}^n} \mu(L) \geqq (2/\omega_n)^{2/n},$$

which completes the proof of 7.2. ∎

These remarks leave many questions unanswered. Does the ratio $\mu(L_n)/n$ tend to a limit as $n \to \infty$? If so, what is the limit? Does the sequence $\mu(L_1), \mu(L_2), \ldots$ increase monotonically? What are the actual values of $\mu(L_9), \mu(L_{10}), \ldots$?

Closely related to the problem of computing $\mu(L_n)$ is the following classical problem. *What is the maximum possible density for a union of non-overlapping balls of fixed radius in the euclidean space* $\mathbf{R}^n$?

Such a union P of non-overlapping balls, all of the same radius, will be called briefly a *packing* of $\mathbf{R}^n$. A packing P is *said* to have *density* ρ if the ratio

$$\operatorname{vol}(P \cap C)/\operatorname{vol}(C),$$

where C denotes a large cube, tends uniformly to the limit ρ as the edge of the cube tends to infinity.

Any lattice $L \subset \mathbf{R}^n$ gives rise to a packing of $\mathbf{R}^n$ as follows. Define the radius r by the equation

$$(2r)^2 = \operatorname*{Min}_{x \in L-0} x \cdot x.$$

Centering a ball of radius r at each lattice point, the interiors of the different balls will evidently be disjoint. The density of this packing is given by the formula

$$\rho = \omega_n r^n / \sqrt{\det L} = \omega_n (\mu(L)/4)^{n/2}.$$

Therefore

$$\mu(L) = 4(\rho/\omega_n)^{2/n}.$$

Note that Minkowski's estimate (1.5) is just the inequality $\rho \leqq 1$.

It was proved by A. Thue that the maximum possible density of a packing of $\mathbf{R}^2$ is equal to the density $\rho=\pi/\sqrt{12}$ associated with the regular hexagonal lattice of Fig. 1. For $n=3$, the face centered cubic lattice gives rise to a packing with density $\pi/\sqrt{18}$. For an arbitrary packing of $\mathbf{R}^3$, according to Rogers, "many mathematicians believe and all physicists know that the density cannot exceed $\pi/\sqrt{18}$". However this has never been proved. Indeed the problem remains unsolved for all $n \geqq 3$. It is noteworthy that Leech and Sloane have recently described examples of non-lattice packings in dimensions 10, 11, 13 which are denser than the packings associated with any known lattices in these dimensions.

An excellent upper bound for the density for any packing of $\mathbf{R}^n$, has been given by Rogers, based on earlier work by Blichfeldt. Let Δ_n be an equilateral n-simplex in euclidean space, with edge length 2. Let B be the subset consisting of all points in Δ_n with distance $\leqq 1$ from some vertex of Δ_n, and let

$$\sigma_n = \mathrm{vol}(B)/\mathrm{vol}(\Delta_n).$$

Then for any packing of R^n which possesses a density ρ, Rogers proves the inequality

$$\rho \leqq \sigma_n.$$

(Here is a table listing some examples, with decimals rounded to the nearest ten-thousandth.)

Dimension n	Lattice	Density of associated packing	Upper bound σ_n
2	regular hexagonal	$0.9069=\pi/\sqrt{12}$	$0.9069=\pi/\sqrt{12}$
3	face centered cubic	0.7405	0.7796
8	Γ_8	0.2537	0.2568
24	Leech lattice	0.0019	0.0025

For any lattice L, it follows that

$$\mu(L)=4(\rho/\omega_n)^{2/n} \leqq 4(\sigma_n/\omega_n)^{2/n}.$$

This represents a substantial improvement over the upper bound $4(1/\omega_n)^{2/n}$ of § 1, as becomes apparent if one uses Rogers asymptotic formula

$$\sigma_n \sim n/e\sqrt{2^n} \quad \text{as } n \to \infty.$$

Numerical values, for $n \leqq 24$, are plotted in Fig. 3. All of the data used in this graph are taken from [Leech] and [Leech, Sloane]. For $n=8$, note that the Rogers upper bound $4(\sigma_8/\omega_8)^{\frac{1}{4}} \approx 2.006$ is so close to $\mu(L_8)=2$ that the two points appear to coincide in the graph.

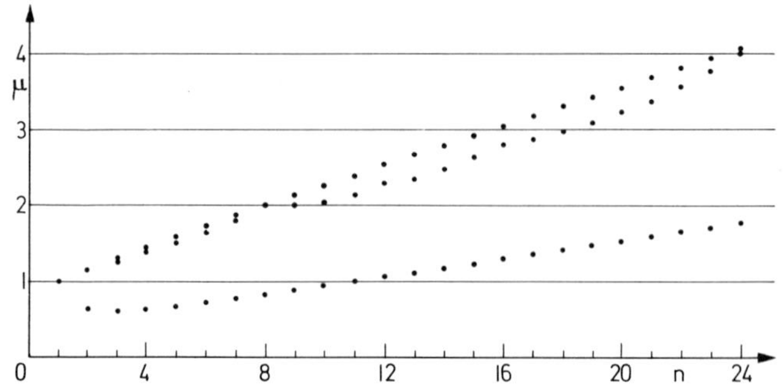

Fig. 3. The middle row of dots gives the largest possible value of $\mu(L) = \mathrm{Min}(x \cdot x)/\sqrt[n]{\det L}$ for a lattice L in each dimension $n \leqq 8$, and the largest known value for each dimension n between 9 and 24. The top row of dots gives the Rogers upper bound $4(\sigma_n/\omega_n)^{2/n}$, and the bottom row gives the crude Minkowski lower bound $(2/\omega_n)^{2/n}$.

To conclude this section we must prove Lemma 7.1. The proof will be based on the following.

(7.4) Lemma. *There exists a constant* $c_n > 0$ *so that every lattice in* $\mathbf{R}^n$ *possesses a basis* $b_1, \ldots, b_n$ *with*

$$|b_i \cdot b_j| \leqq c_n \det(L)/(\underset{L-0}{\mathrm{Min}}\, x \cdot x)^{n-1}$$

for all i, j.

In fact we will show that a suitable constant c_n can be defined inductively by the formula

(5) $$c_n = (\tfrac{4}{3})^{n-2} c_{n-1} + 4^n \omega_n^{-2},$$

with $c_1 = 1$.

To begin the proof of (7.4), choose a vector $b_n \neq 0$ in L with

(6) $$b_n \cdot b_n = \underset{x \in L-0}{\mathrm{Min}}\, x \cdot x.$$

Then

(7) $$b_n \cdot b_n \leqq 4\omega_n^{-2/n} (\det L)^{1/n}$$

by 1.5 or 7.2. Dividing the n-th power of (7) by the $(n-1)$-st power of (6), we obtain the inequality

(8) $$b_n \cdot b_n \leqq (4^n \omega_n^{-2}) \det L/(\underset{L-0}{\mathrm{Min}}\, x \cdot x)^{n-1},$$

where the coefficient $4^n \omega_n^{-2}$ evidently satisfies $4^n \omega_n^{-2} \leqq c_n$.

This completes the proof when $n=1$. Suppose then that $n \geqq 2$. Let L' denote the image of L under orthogonal projection from $\mathbf{R}^n$ to the hyperplane $(b_n)^{\perp}$. Then every vector $x' \in L'$ can be expressed as a difference

$$x' = x - \lambda b_n$$

with $x \in L$, and we can choose x so that $|\lambda| \leqq \frac{1}{2}$. If $x' \neq 0$, it follows that

$$b_n \cdot b_n \leqq x \cdot x = x' \cdot x' + \lambda^2 b_n \cdot b_n \leqq x' \cdot x' + \tfrac{1}{4} b_n \cdot b_n,$$

hence

$$x' \cdot x' \geqq \tfrac{3}{4} b_n \cdot b_n = \tfrac{3}{4} \operatorname*{Min}_{L-0} x \cdot x.$$

Now, using the formulas

$$\operatorname*{Min}_{L'-0} x' \cdot x' \geqq \tfrac{3}{4} \operatorname*{Min}_{L-0} x \cdot x$$

and

$$\det(L') = \det(L)/b_n \cdot b_n,$$

it follows inductively that L' possesses a basis $b'_1, \ldots, b'_{n-1}$ with

(9) $b'_i \cdot b'_i \leqq c_{n-1} \det(L')/(\operatorname*{Min}_{L'-0} x' \cdot x')^{n-2} \leqq (\frac{4}{3})^{n-2} c_{n-1} \det(L)/(\operatorname*{Min}_{L-0} x \cdot x)^{n-1}$.

Setting $b'_i = b_i - \lambda_i b_n$ with $|\lambda_i| < 1$, we obtain the required elements $b_1, \ldots, b_{n-1}$, satisfying

$$b_i \cdot b_i < b'_i \cdot b'_i + b_n \cdot b_n \leqq c_n \det L/(\operatorname*{Min}_{L-0} x \cdot x)^{n-1}$$

by (5), (8), (9). Evidently $b_1, \ldots, b_n$ form a basis for L. Since the required upper bound for $|b_i \cdot b_j|$, with $i \neq j$, now follows from the Schwarz inequality, this completes the proof of (7.4). □

Proof of Lemma (7.1). Let K be the set consisting of all symmetric $n \times n$ matrices $A = (a_{ij})$ of real numbers which satisfy the inequality

(10) $$\sum a_{ij} \zeta_i \zeta_j \geqq 1$$

for every n-tuple $(\zeta_1, \ldots, \zeta_n)$ of integers, not all zero. Evidently K can be thought of as a closed and convex subset of real $n(n+1)/2$-space.

Note that every matrix in K is positive definite. In fact it follows from 1.5 or 7.2 that every positive definite matrix A in K satisfies the inequality

$$\det(A) \geqq \omega_n^2/4^n > 0.$$

It follows easily that the subset of K consisting of positive definite matrices is not only open, but also closed. Hence this subset equals K.

We will prove that the determinant function $A \mapsto \det(A)$ from K to $\mathbf{R}$ actually attains a minimum value at some point $A_0 \in K$. Let $K_0 \subset K$ be the compact subset consisting of all matrices $A = (a_{ij})$ in K whose entries a_{ij} satisfy

$$|a_{ij}| \leqq c_n.$$

It follows from 7.4 that for every matrix $A \in K$ with $\det(A) \leqq 1$ there exists an invertible integer matrix $P \in \mathrm{GL}(n, \mathbf{Z})$ so that $PAP^t \in K_0$. Evidently

$$\det(PAP^t) = \det(A).$$

But the determinant function restricted to the compact set K_0 certainly attains a minimum value. Therefore we can choose $A_0 \in K_0$ with $\det(A_0) \leqq \det(A)$ for all $A \in K_0$, and therefore for all $A \in K$.

If L_n is a lattice with inner product matrix A_0, then evidently it follows that

$$\mu(L_n) = 1/\sqrt[n]{\det A_0}$$

is equal to $\operatorname*{Max}_{L \subset \mathbf{R}^n} \mu(L)$.

Now we must prove that A_0 is a matrix of rational numbers. More generally, following Korkine and Zolotareff, a matrix $A_0 \in K$ is called *extreme* if the function det: $K \to \mathbf{R}$ has a local minimum at A_0. *Note that an extreme matrix A_0 cannot lie in the interior of a line segment of matrices*

$$A_\xi = A_0 + \xi B, \qquad -\varepsilon < \xi < \varepsilon,$$

all belonging to K, with $B \neq 0$. (In the language of convex set theory, an extreme matrix is necessarily an "extreme point" of the convex set K.) For choosing a real matrix P with $PA_0P^t = I$, and setting $PBP^t = C = (c_{ij})$, a short computation shows that

$$\det(A_\xi)/\det(A_0) = \det(I + \xi C) = \tfrac{1}{2} + \tfrac{1}{2}\Bigl(1 + \xi \sum_i c_{ii}\Bigr)^2 - \tfrac{1}{2}\xi^2 \sum_i \sum_j c_{ij}^2 + \cdots$$

omitting terms in ξ^3 and higher powers. Since $C \neq 0$, it follows easily that there exist arbitrarily small values of ξ with $\det(A_\xi) < \det(A_0)$.

If A_0 is an extreme matrix, let $\pm z_{(1)}, \ldots, \pm z_{(N)}$ be a complete list of all n-tuples of integers (thought of as $1 \times n$ matrices) which satisfy the matrix equation

$$z_{(i)} A_0 z_{(i)}^t = 1. \tag{11}$$

We will consider (11) as a collection of N linear equations for the $n(n+1)/2$ entries of the matrix A_0. *We claim that these equations can be solved uniquely for A_0.* For otherwise, if the solution were not unique, there would exist a one-parameter family of symmetric matrices $A_\xi = A_0 + \xi B$, all satisfying the equations (11). We will prove that $A_\xi \in K$ for ξ sufficiently small. If z is a non-zero n-tuple of integers, then either $zA_0z^t = 1$ and hence $zA_\xi z^t = 1$ also, or $zA_0z^t \geqq c$ for some constant $c > 1$. In the latter case, choosing a constant $\varepsilon > 0$ so that

$$|xBx^t| \leqq \varepsilon^{-1} x A_0 x^t$$

for all $x \in \mathbf{R}^n$, it follows that

$$z A_\xi z^t \geqq z A_0 z^t - \varepsilon^{-1} |\xi| z A_0 z^t > 1$$

whenever $|\xi| < (1 - c^{-1})\varepsilon$. Thus $A_\xi \in K$ for $|\xi| < (1 - c^{-1})\varepsilon$, which contradicts the hypothesis that A_0 is extreme.

Since the unique solution of rational linear equations is necessarily rational, this shows that A_0 is a matrix of rational numbers, and completes the proof of (7.1). ∎

§ 8. Sums of Two and Four Squares

This section will use Minkowski's convex body theorem to prove two venerable theorems. The first was stated by Fermat in the seventeenth century, and proved by Euler in the eighteenth century.

(8.1) Theorem. *If p is a prime number of the form $4k+1$, then the equation $a^2 + b^2 = p$ has a solution with $a, b \in \mathbf{Z}$.*

In terms of the ring $\mathbf{Z}[i]$ of Gaussian integers, this theorem states that every such prime splits as a product $(a + b\,i)(a - b\,i)$.

Proof. Let p be any odd prime. First note that the congruence $u^2 \equiv -1 \pmod{p}$ has a solution $u \in \mathbf{Z}$ if and only if $p \equiv 1 \pmod{4}$. For the group $(\mathbf{Z}/p\mathbf{Z})^\bullet$ of relatively prime residue classes modulo p is cyclic of order $p-1$. The image of -1 in $(\mathbf{Z}/p\mathbf{Z})^\bullet$ is the unique element of order 2. Evidently there exists an element of order 4 in this group if and only if 4 divides $p-1$, that is, if and only if $p \equiv 1 \pmod{4}$.

Suppose now that p is congruent to 1 mod 4, and choose some integer u satisfying the congruence $u^2 \equiv -1 \pmod{p}$. Fixing u, let $L \subset \mathbf{Z} \oplus \mathbf{Z}$ be the lattice in $\mathbf{R}^2$ consisting of all pairs (a, b) of integers satisfying

$$b \equiv u\,a \pmod{p}.$$

Then L is a subgroup of index p in $\mathbf{Z}^2$, hence the volume of a fundamental domain for L is p. Therefore by (1.5) there exists a lattice point $x \neq 0$ with

$$x \cdot x \leqq 4p/\pi.$$

Setting $x = (a, b)$, it follows that

$$0 < a^2 + b^2 \leqq 4p/\pi < 2p.$$

But modulo p we have

$$a^2 + b^2 \equiv a^2 + (u\,a)^2 \equiv (1 + u^2)\,a^2 \equiv 0.$$

Therefore $a^2 + b^2$ must be precisely equal to p. This completes the proof. ∎

(8.2) Corollary. *The subset of* $\mathbf{Q}^{\bullet}$ *consisting of all non-zero rationals which can be expressed as the sum of two squares is a free abelian multiplicative group, with basis* $2, 3^2, 5, 7^2, 11^2, 13, 17, \ldots$.

(Compare Chapter III, § 4.4.)

Proof. First consider any equation of the form $\alpha^2+\beta^2=\gamma\neq 0$ in $\mathbf{Q}$. After multiplying by a large square, we may assume that $\alpha, \beta, \gamma \in \mathbf{Z}$. We must show that any odd prime p which divides γ to an odd power must be of the form $4k+1$. Let p^{2i} be the highest power of p dividing both α^2 and β^2. Then evidently

$$(\alpha/p^i)^2 \equiv -(\beta/p^i)^2 \not\equiv 0 \pmod{p}.$$

Hence -1 is a quadratic residue modulo p, and by the first remark in the proof of (8.1) it follows that $p\equiv 1 \pmod 4$. Thus if $\gamma=\alpha^2+\beta^2$ then γ must belong to the multiplicative group generated by $2, 3^2, 5, 7^2, \ldots$. Conversely, using the identity

$$(a^2+b^2)(c^2+d^2)=(a\,c-b\,d)^2+(a\,d+b\,c)^2$$

it follows easily that every element of this multiplicative group is a sum of squares. □

Here is an exercise for the reader. Prove that an integer is a sum of two rational squares if and only if it is a sum of two integer squares.

Our second theorem was stated by Bachet de Méziriac in the seventeenth century, and proved by Lagrange in the eighteenth.

(8.3) Theorem. *Every positive integer is a sum of four squares.*

It follows easily that every positive rational is a sum of four rational squares. (This can be interpreted as a special case of Meyer's theorem, corresponding to inner product spaces of the form $\langle 1\rangle\oplus\langle 1\rangle\oplus\langle 1\rangle\oplus\langle 1\rangle\oplus\langle -r\rangle$.)

Proof. Let p be any odd prime. By Lemma 3.3, we see that the congruence

$$u^2+v^2+1\equiv 0 \bmod p$$

has a solution $u, v\in\mathbf{Z}$. (The "shoe box principal" shows that one of the $(p+1)/2$ distinct values of u^2 modulo p must coincide with one of the $(p+1)/2$ distinct values of $-1-v^2$.) Fixing u and v, let $L\subset\mathbf{Z}\oplus\mathbf{Z}\oplus\mathbf{Z}\oplus\mathbf{Z}$ be the lattice consisting of all 4-tuples (a, b, c, d) with

$$c\equiv u\,a+v\,b, \quad d\equiv u\,b-v\,a \pmod{p}.$$

Then L is a subgroup of index p^2 in $\mathbf{Z}^4$, so it follows that the volume of a fundamental domain for L is p^2. Therefore, by (1.5), there exists a lattice point $x\neq 0$ with

$$x\cdot x\leqq 4\sqrt{p^2/\omega_4}=4p\sqrt{2}/\pi.$$

Setting $x=(a, b, c, d)$, it follows that

$$0<a^2+b^2+c^2+d^2 \leqq 4p\sqrt{2}/\pi < 2p.$$

But, working modulo p, we have

$$\begin{aligned} a^2+b^2+c^2+d^2 &\equiv a^2+b^2+(u\,a+v\,b)^2+(u\,b-v\,a)^2 \\ &=(1+u^2+v^2)\,a^2+(1+u^2+v^2)\,b^2 \\ &\equiv 0 \pmod{p}. \end{aligned}$$

Therefore $a^2+b^2+c^2+d^2$ must be precisely equal to p.

To extend this result to an arbitrary product of primes, we note that $a^2+b^2+c^2+d^2$ is the norm of the quaternion $a+b\,i+c\,j+d\,k$. Since the norm function from the quaternions to the reals is multiplicative, it follows that any product $(a^2+b^2+c^2+d^2)\,(A^2+B^2+C^2+D^2)$ of sums of four squares can itself be expressed as a sum

$$\begin{aligned} &(aA-bB-cC-dD)^2+(aB+bA+cD-dC)^2 \\ &\quad+(aC-bD+cA+dB)^2+(aD+bC-cB+dA)^2 \end{aligned}$$

of four squares. Evidently this completes the proof. □

For a characterization of sums of three squares, see for example [Serre, p. 79].

§ 9. A Theorem of Siegel

This section will describe a basic formula due to C. L. Siegel, concerning positive definite bilinear forms over **Z**. No proof will be given, but some interesting applications will be described in detail. For proofs the reader is referred to [Siegel], [Weil], [Mars].

One of the classical problems of number theory is that of giving a satisfactory formula for the number of solutions to a quadratic equation in several integer variables. For example, given a positive integer k, how many n-tuples $x_1, \ldots, x_n \in \mathbf{Z}$ satisfy the equation $x_1^2+\cdots+x_n^2=k$? In 1881 the Paris Academy announced that its Grand Prize would be awarded to the author of the best paper on the topic "Théorie de la décomposition des nombres entiers en une somme de cinq carrés." The French press was outraged, two years later, when half[3a] of the prize was awarded to a teen-aged German student who, because of time pressure, had not even followed the rules and submitted his manuscript in the French language. However the French Academy stood firm. In his prize winning paper, Hermann Minkowski laid the groundwork for our

[3a] The other prize winner was H. J. S. Smith of Oxford, who had actually published a solution to the stated problem years earlier.

present theory of quadratic forms over the integers and over the rational numbers.

This section will be concerned with Siegel's formula for the "average" number of solutions to a quadratic equation. In order to state the formula, a number of definitions will be needed.

Let X and Y be topological spaces, each provided with a measure on the σ-ring generated by open sets. The measure of a set U will be denoted by $\mathrm{vol}(U)$. Given any continuous map $f\colon X\to Y$, the density of solutions to the equation $f(x)=y_0$ can be measured as follows. Consider small neighborhoods U of y_0, and form the limit

$$\lim_{U\to y_0} \mathrm{vol}\, f^{-1}(U)/\mathrm{vol}(U).$$

If this limit exists, then it will be called the *density* of f^{-1} at y_0, and denoted by $Df^{-1}(y_0)$. If this density is continuous as a function of y_0, then evidently

$$\int_U Df^{-1}(y)\,dy=\mathrm{vol}\, f^{-1}(U)$$

for any measurable set $U\subset Y$. (This is a topological version of the Radon-Nikodym theorem.)

Example 1. Let $f\colon \mathbf{R}^n\to\mathbf{R}$ be the quadratic function $f(x_1,\dots,x_n)=x_1^2+\cdots+x_n^2$. Using the usual Lebesgue measure on $\mathbf{R}^n$ and on $\mathbf{R}$, we see that

$$\int_0^y Df^{-1}(\eta)\,d\eta=\mathrm{vol}\, f^{-1}[0,y]$$

is equal to the volume of a ball of radius $\sqrt{y}$ for $y>0$. Therefore $Df^{-1}(y)$ is equal to the derivative

$$\frac{d}{dy}(\omega_n\sqrt{y^n})=\tfrac{1}{2}\,n\,\omega_n\,y^{n/2-1}.$$

For example

$$\begin{aligned} Df^{-1}(y)&=1/\sqrt{y} && \text{if } n=1,\\ &=\pi && \text{if } n=2,\\ &=2\pi\sqrt{y} && \text{if } n=3,\\ &=\pi^2 y && \text{if } n=4, \end{aligned}$$

and so on, provided that $y>0$. Evidently $Df^{-1}(y)=0$ for $y<0$. The function Df^{-1} is continuous at 0 for $n\geqq 3$, but is discontinuous for $n=1,2$.

Example 2. Let $\mathbf{Z}_p$ denote the ring of p-adic integers. This ring has a canonical Haar measure, in which the volume of each of the p^k distinct residue classes modulo p^k is equal to p^{-k}. The n-fold cartesian product $\mathbf{Z}_p \times \cdots \times \mathbf{Z}_p$ has a corresponding product Haar measure. Defining $f_p\colon \mathbf{Z}_p \times \cdots \times \mathbf{Z}_p \to \mathbf{Z}_p$ by $f_p(x_1, \ldots, x_n) = x_1^2 + \cdots + x_n^2$, we can again form the density $D f_p^{-1}\colon \mathbf{Z}_p \to \mathbf{R}$.

Let u be any p-adic unit. Then the transformation $(x_1, \ldots, x_n) \mapsto (u x_1, \ldots, u x_n)$ is continuous and volume preserving. Since

$$f_p(u x_1, \ldots, u x_n) = u^2 f_p(x_1, \ldots, x_n),$$

it follows that $D f_p^{-1}(u^2 y)$ is equal to $D f_p^{-1}(y)$. *In particular, for every* $y \neq 0$ *in* $\mathbf{Z}_p$ *the function* $D f_p^{-1}$ *is constant throughout a neighborhood of* y.

Here is an explicit formula for $D f_p^{-1}$. To simplify the discussion we consider only the case n even, say $n = 2m$. Recall that every element $y \neq 0$ in $\mathbf{Z}_p$ can be written uniquely as a product $p^v u$ where v is a non-negative integer and u is a p-adic unit.

(9.1) Lemma. *Let* $f_p\colon \mathbf{Z}_p \times \cdots \times \mathbf{Z}_p \to \mathbf{Z}_p$ *be the function* $f_p(x_1, \ldots, x_{2m}) = x_1^2 + \cdots + x_{2m}^2$ *of* $2m$ *p-adic variables. Let* $r = p^{1-m}$, *and for* p *odd let* $\varepsilon = \left(\frac{-1}{p}\right)^m$ *so that* ε *equals* $+1$ *or* -1 *according as* $p^m \equiv 1$ *or* $p^m \equiv -1$ (mod 4). *If* p *is odd then the associated density function* $D f_p^{-1}(p^v u)$ *is equal to*

$$(1 - \varepsilon r/p)(1 + \varepsilon r + \varepsilon^2 r^2 + \cdots + \varepsilon^v r^v).$$

For $p = 2$ *and* m *even it is equal to*

$$1 + (-1)^{m/2}(r + r^2 + \cdots + r^{v-1} - r^v),$$

while for $p = 2$ *and* m *odd it is equal to either* $1 + r^{v+1}$ *or* $1 - r^{v+1}$ *according as* $m \equiv u$ *or* $m \equiv -u$ (mod 4).

The proof will be given at the end of this section. The formulas for n odd are slightly more complicated.

Note that $D f_p^{-1}(y)$ tends to 1 as $m \to \infty$. Furthermore, if $m \geqq 2$, note that $D f_p^{-1}(y)$ tends to 1 uniformly as $p \to \infty$.

It will be convenient to unify our notations by declaring that the symbol $\mathbf{Z}_\infty$ shall stand for the real numbers $\mathbf{R}$. Thus we have a canonical embedding $\mathbf{Z} \to \mathbf{Z}_p$ for $p = 2, 3, 5, 7, \ldots, \infty$.

Definition (Gauss, Minkowski). Two bilinear form spaces X and Y over the integers belong to the same *genus* if the induced bilinear form space $X \otimes \mathbf{Z}_p$ over $\mathbf{Z}_p$ is isomorphic to $Y \otimes \mathbf{Z}_p$ for every $p = 2, 3, \ldots, \infty$.

If X and Y have the same genus, then it is not difficult to verify that they also have the same determinant. Hence, using (1.6), we see that *each genus contains only finitely many distinct spaces, up to isomorphism.*

Example 3. The two bilinear form spaces $\langle 5\rangle \oplus \langle 11\rangle$ and $\langle 1\rangle \oplus \langle 55\rangle$ are both positive definite, and it is not difficult to show that the injection $\mathbf{Z} \to \mathbf{Z}_p$ carries them into isomorphic spaces for every finite prime p. Hence these two spaces belong to the same genus. They are not isomorphic, since the equation $5x^2 + 11y^2 = 1$ has no integer solution.

Similarly the two inner product spaces $\langle 1\rangle \oplus \Gamma_8$ and $\langle 1\rangle \oplus \cdots \oplus \langle 1\rangle$ of rank 9 belong to the same genus, but are not isomorphic.

We are now ready to state one form of Siegel's theorem. Let X be a positive definite bilinear form space of rank $n \geqq 2$ over $\mathbf{Z}$.

Definition. For each integer k, let $r_X(k)$ denote the number of distinct elements $x \in X$ satisfying the equation $x \cdot x = k$.

For each p we can form the induced space $X \otimes \mathbf{Z}_p$ over $\mathbf{Z}_p$. Let

$$f_p\colon\ X \otimes \mathbf{Z}_p \to \mathbf{Z}_p$$

denote the quadratic function $f_p(\xi) = \xi \cdot \xi$.

(9.2) Siegel's theorem (preliminary version). *If the genus of X contains only one isomorphism class, then*

$$r_X(k) = \varepsilon \prod_{p=2,3,\ldots,\infty} D f_p^{-1}(k)$$

for every integer $k \neq 0$, where the coefficient ε is defined to be $\frac{1}{2}$ or 1 according as $n=2$ or $n>2$.

For the proof, which is not easy, we refer to [Siegel, pp. 326–366]. This theorem remains true in the indefinite case, but is not interesting since both sides of the equation are usually infinite.

The product on the right is absolutely convergent if $n \geqq 4$. In the cases $n = 2, 3$, care must be taken to multiply the factors in the usual order.

Example 4. The inner product space $\langle 1\rangle \oplus \langle 1\rangle \oplus \cdots \oplus \langle 1\rangle$ of rank 8 satisfies the hypothesis of (9.2). Let $v_p = v_p(k)$ denote the exponent of the highest power of p dividing k. Then, by (9.1),

$$D f_p^{-1}(k) = (1 - p^{-4})(1 + p^{-3} + p^{-6} + \cdots + p^{-3v_p})$$

for p odd, and

$$D f_2^{-1}(k) = 1 + 2^{-3} + 2^{-6} + \cdots + 2^{-3(v_2-1)} - 2^{-3v_2}$$

for $p = 2$. (If $v_2 = 0$, this formula reads $D f_2^{-1}(k) = 1$.) Furthermore

$$D f_\infty^{-1}(k) = 4\omega_8 k^3 = \pi^4 k^3/6$$

by Example 1. Taking the product over $p = 2, 3, \ldots, \infty$, we obtain an infinite product

$$C = \tfrac{1}{6}\pi^4 (1 - 3^{-4})(1 - 5^{-4})(1 - 7^{-4})\ldots$$

which does not depend on k, multiplied by a finite product

$$k^3 \prod_{p|k} (1+p^{-3}+p^{-6}+\cdots \pm p^{-3v_p})$$

which depends explicitly on k. Multiplying out this last expression, we obtain

$$k^3 \sum_{d|k} \pm d^{-3} = \sum_{d|k} \pm d^3,$$

where the sign of the summand d^3 turns out to be $(-1)^{d+k}$.

As for the constant C, evaluating $r_X(k)$ at $k=1$ we see that $C=16$. Or alternatively, in terms of the Riemann zeta function we can write

$$C = \tfrac{1}{6}\pi^4 (1-2^{-4})^{-1} \prod_{p \text{ finite}} (1-p^{-4}) = \tfrac{1}{6}\pi^4 \tfrac{16}{15} \zeta(4)^{-1}.$$

Substituting the known identity

$$\zeta(4) = \pi^4/90$$

(see for example [Serre, p. 148]), we again obtain $C=16$. Thus we have derived the following.

Formula of Jacobi. *For any positive integer k, the number of representations of k as a sum of eight squares is equal to* $16 \sum_{d|k} (-1)^{d+k} d^3$.

We leave the corresponding formulas for sums of two, four, or six squares as exercises for the reader.

Now consider an arbitrary genus G of positive definite bilinear form spaces. Then G contains say g distinct isomorphism classes, where g is a positive integer. Let $X_1, \ldots, X_g$ be representatives for these various isomorphism classes. Following Eisenstein, we weight the various isomorphism classes according to their lack of symmetry. More precisely, let $|O(X_i)|$ denote the order of the orthogonal group consisting of all automorphisms of X_i. Defining

$$w_i = |O(X_i)|^{-1} \Big/ \sum_{j=1}^{g} |O(X_j)|^{-1},$$

we obtain positive rational numbers $w_1, \ldots, w_g$ with $w_1 + \cdots + w_g = 1$.

Let $f_p\colon X_1 \otimes \mathbf{Z}_p \to \mathbf{Z}_p$ be the quadratic functions $f_p(x) = x \cdot x$, as above.

(9.3) Siegel's theorem (second version). *For any $k \neq 0$, the weighted average $w_1 r_{X_1}(k) + \cdots + w_g r_{X_g}(k)$ is equal to $\varepsilon \prod_{p=2,3,\ldots,\infty} Df_p^{-1}(k)$, where ε equals $\frac{1}{2}$ or 1 according as the rank is 2 or greater than 2.*

Suppose for example that X_1 is the inner product space $\langle 1\rangle \oplus \cdots \oplus \langle 1\rangle$ of rank n. Then the genus G consists of all positive definite inner product spaces of type I and rank n. We denote this genus by the symbol I_n.

If n is large, then the finite primes contribute very little to the product $\prod D f_p^{-1}(k)$. The factor

$$D f_\infty^{-1}(k) = \tfrac{1}{2}\, n\, \omega_n\, k^{n/2-1}$$

is dominant. Here is a very crude estimate.

(9.4) Lemma. *If $f_p(x_1, \ldots, x_n) = \sum x_i^2$, with $n \geqq 8$, then the product over all finite primes of $D f_p^{-1}(k)$ lies between $\frac{5}{6}$ and $\frac{6}{5}$.*

The proof will be given later.

Hence the average $w_1 r_{X_1}(k) + \cdots + w_g r_{X_g}(k)$ lies between $\frac{5}{12} n\, \omega_n\, k^{n/2-1}$ and $\frac{6}{10} n\, \omega_n\, k^{n/2-1}$.

J.H. Conway and J. Thompson have pointed out (unpublished) that Siegel's theorem can be used to prove an analogue of the Minkowski-Hlawka Theorem (7.2) for inner product spaces (i.e., self-dual lattices) over **Z**.

For each $n > 0$ let $k(n)$ denote the closest integer to $(\frac{5}{3}\omega_n^{-1})^{2/n}$. Clearly this integer $k(n)$ is asymptotic to $\omega_n^{-2/n} \sim n/2\pi e$ as $n \to \infty$.

(9.5) Theorem (Conway, Thompson). *For any dimension n there exists a positive definite inner product space X of type I and rank n with*

$$\operatorname*{Min}_{x \in X - 0} x \cdot x \geqq k(n).$$

Proof. Let $k = k(n)$. We may assume that $n \geqq 8$, since $k \leqq 1$ for smaller values of n. Hence Lemma (9.4) applies. If $X_1, \ldots, X_g$ are the distinct inner product spaces in the genus I_n, then $w_1 r_{X_1}(j) + \cdots + w_g r_{X_g}(j) < \frac{6}{10} n\, \omega_n\, j^{n/2-1}$. Summing over $j = 1, 2, \ldots, k-1$, and using the inequality

$$j^m \leqq \int_{m-\frac{1}{2}}^{m+\frac{1}{2}} t^m\, dt$$

which is valid for $m \geqq 1$, we see that the weighted average

$$\sum_{i=1}^{g} w_i \big(r_{X_i}(1) + r_{X_i}(2) + \cdots + r_{X_i}(k-1)\big)$$

is less than

$$\tfrac{6}{10}\, n\, \omega_n \int_0^{k-\frac{1}{2}} t^{n/2-1}\, dt = \tfrac{6}{5}\, \omega_n (k - \tfrac{1}{2})^{n/2}.$$

But $k = k(n)$ was chosen so that

$$k - \tfrac{1}{2} \leqq (\tfrac{5}{3}\, \omega_n^{-1})^{2/n}.$$

Therefore this upper bound is less than or equal to $\frac{6}{5}\omega_n \frac{5}{3}\omega_n^{-1} = 2$. Since the weighted average is less than 2, there must exist some particular inner product space $X = X_i$ so that

$$r_X(1) + r_X(2) + \cdots + r_X(k-1) < 2.$$

Each $r_X(j)$ is an even, non-negative integer, so this inequality can be satisfied only if

$$r_X(1) = r_X(2) = \cdots = r_X(k-1) = 0.$$

Hence

$$\min_{x \in X - 0} x \cdot x \geqq k = k(n),$$

which completes the proof. ▯

Here are some numerical values. Computation shows that $k(20) = 2$, $k(37) = 3$, $k(54) = 4$, $k(71) = 5$. It follows for example that for all $n \geqq 37$ the genus I_n contains a space X with $\min_{X-0} x \cdot x \geqq 3$.

A similar argument can be given for spaces of type II. (Compare [Serre, p. 176].) In this case, for $n \equiv 0 \bmod 8$, there exists a space X of type II with $\min_{X-0} x \cdot x$ greater than or equal to the closest even integer to $(\frac{5}{3}\omega_n^{-1})^{2/n}$. For example if $n = 8m \geqq 48$ then there exists a space of rank n and type II with $\min_{X-0} x \cdot x \geqq 4$.

An interesting application of (9.5) has been pointed out by R. Steinberg (unpublished). For any lattice L in $\mathbf{R}^n$ let

$$m(L) = \min_{x \in L - 0} x \cdot x.$$

Given two lattices $L \subset \mathbf{R}^n$ and $L' \subset \mathbf{R}^k$ we can form the tensor product

$$L \otimes L' \subset \mathbf{R}^n \otimes \mathbf{R}^k \cong \mathbf{R}^{nk}.$$

It is clear that

$$m(L \otimes L') \leqq m(L)\, m(L'),$$

and one would be tempted to conjecture that equality holds.

(9.6) Theorem (Steinberg). *If L' is the lattice Γ_8 of § 6, then the equality*

$$m(L \otimes \Gamma_8) = m(L)\, m(\Gamma_8) = 2m(L)$$

is valid for any lattice $L \subset \mathbf{R}^n$. On the other hand for any dimension $n \geqq 292$ there exist self dual lattices L and L' in $\mathbf{R}^n$ so that

$$m(L \otimes L') < m(L)\, m(L').$$

To prove the first statement, we note that $L \otimes \Gamma_8$ can be described as the set of all elements $u_1 \otimes e_1 + \cdots + u_8 \otimes e_8$ in $\mathbf{R}^n \otimes \mathbf{R}^8$ for which

$$u_i \in \tfrac{1}{2} L, \quad u_1 \equiv u_2 \equiv \cdots \equiv u_8 \bmod L, \quad \text{and} \quad u_1 + \cdots + u_8 \in 2L.$$

Suppose that $x = \sum u_i \otimes e_i$ is a non-zero element of $L \otimes \Gamma_8$. If $u_1, \ldots, u_8 \in 2L$, then

$$x \cdot x = \sum u_i \cdot u_i \geqq m(2L) = 4m(L).$$

If one of the u_i belongs to L but not to $2L$, then some other, say u_j, must also belong to L but not to $2L$; hence $u_i \neq 0$, $u_j \neq 0$, and

$$x \cdot x \geqq u_i\, u_i + u_j \cdot u_j \geqq 2m(L).$$

Finally, if some u_i does not belong to L, then $u_j \neq 0$ for all j, and therefore

$$x \cdot x \geqq 8m(\tfrac{1}{2} L) = 2m(L).$$

Thus $m(L \otimes \Gamma_8) \geqq 2m(L)$, and it clearly follows that equality holds.

As an example, consider the tensor product of k copies of Γ_8. It evidently follows inductively that $m(\Gamma_8 \otimes \cdots \otimes \Gamma_8) = 2^k$.

To prove the second statement, let L be any lattice in $\mathbf{R}^n$ and let $L^{\#}$ denote the dual lattice consisting of all $x \in \mathbf{R}^n$ with $x \cdot L \subset \mathbf{Z}$. The tensor product $L \otimes L^{\#}$ is canonically isomorphic to $\operatorname{Hom}(L, L)$, so $L \otimes L^{\#}$ contains a distinguished element e corresponding to the identity map of L. In terms of a basis $b_1, \ldots, b_n$ and dual basis $b_1^{\#}, \ldots, b_n^{\#}$ one has

$$e = b_1 \otimes b_1^{\#} + \cdots + b_n \otimes b_n^{\#}.$$

A short computation shows that $e \cdot e = n$. Therefore

$$m(L \otimes L^{\#}) \leqq n.$$

Now let us apply (9.5). For any n there exists an inner product space (that is a self-dual lattice) $L = L^{\#}$ with

$$m(L) \geqq k(n) \sim n/2\pi e.$$

For large n this is evidently greater than $\sqrt{n}$. In fact, if $n > (2\pi e)^2 = 291.708\ldots$ then computation shows that $k(n) > \sqrt{n}$. Thus, choosing L with

$$m(L) = m(L^{\#}) > \sqrt{n},$$

it follows that

$$m(L)\, m(L^{\#}) > n \geqq m(L \otimes L^{\#}),$$

which completes the proof. □

Now let us describe a still more general version of Siegel's formula. Instead of solving a single equation $x \cdot x = k$, suppose that we try to solve a collection of $t(t+1)/2$ simultaneous equations of the form

$$x_i \cdot x_j = K_{ij}$$

where $K=(K_{ij})$ is a fixed symmetric $t\times t$ matrix of integers and $x_1, \ldots, x_t$ are unknown elements of X. We assume that $1\leqq t\leqq n$. Let $r_X(K)$ denote the number of solutions. In other words $r_X(K)$ is the number of elements in $f^{-1}(K)$, where

$$f: X\times\cdots\times X\to(\text{additive group of } t\times t \text{ symmetric matrices})$$

is the quadratic function $f(x_1, \ldots, x_t)=(x_i\cdot x_j)$. Just as above, we can tensor everything with $\mathbf{Z}_p$, and form a corresponding p-adic function f_p.

(9.7) Siegel's theorem (final version). *The weighted average*

$$w_1 r_{X_1}(K)+\cdots+w_g r_{X_g}(K),$$

where $X_1, \ldots, X_g$ represent the distinct isomorphism classes in the genus of $X=X_1$, is equal to

$$(\varepsilon_{n-t-1}/\varepsilon_{n-1})\prod_{p=2,3,\ldots,\infty}\varepsilon_{n-t}\, D f_p^{-1}(K),$$

where the coefficient ε_i is equal to $\frac{1}{2}$ if $i=0$, and to 1 if $i\neq 0$.

Evidently this reduces to the previous theorem in the case $n>t=1$.

By way of contrast, let us look at the case $n=t>1$. Choosing a basis $b_1, \ldots, b_n$ for X_1, suppose that K_1 is the matrix $(b_i\cdot b_j)$. Then $r_{X_1}(K_1)$ is evidently equal to the number $|O(X_1)|$ of automorphisms of X_1, while $r_{X_i}(K_1)=0$ for $i>1$. Thus the weighted average $\sum w_i r_{X_i}(K_1)$ reduces to $w_1|O(X_1)|$. But by definition

$$w_1=|O(X_1)|^{-1}/\sum_j |O(X_j)|^{-1}.$$

Therefore the weighted average is equal to the expression

$$\left(\sum_{j=1}^{g}|O(X_j)|^{-1}\right)^{-1}$$

which depends only on the genus G. Classically the reciprocal

$$M(G)=\sum_{j=1}^{g}|O(X_j)|^{-1}$$

is called the *mass* associated with the genus G. Thus we have a more or less effective formula $M(G)=\prod_{p=2,\ldots,\infty}\left(\frac{1}{2}Df_p^{-1}(K_1)\right)^{-1}$ for computing the mass associated with any genus of positive definite spaces over $\mathbf{Z}$.

Note the basic inequality $g\geqq 2M(G)$, where g is the number of distinct isomorphism classes in G. This is evident since each automorphism group $O(X)$ contains at least two distinct elements (namely 1 and -1).

As an example, consider the genus I_n consisting of all positive definite inner product spaces of type I and rank n. Then the function $M(I_n)$ is plotted in Fig. 4, on a highly condensed logarithmic scale. For small

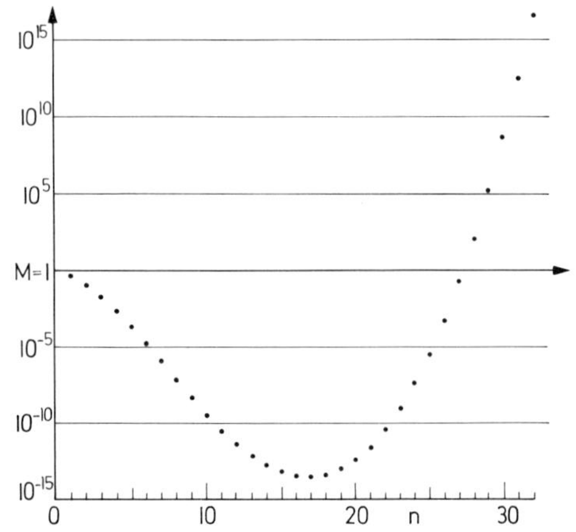

Fig. 4. The mass $M(I_n)$ of the genus I_n plotted on a logarithmic scale as a function of n.

values of n the mass $M(I_n)$ is very close to zero. For example if $n \leqq 8$ then $M(I_n)$ is just the reciprocal of the number $n!\,2^n$ of automorphisms of the n-fold sum $\langle 1\rangle \oplus \cdots \oplus \langle 1\rangle$. But for larger values of n the mass $M(I_n)$ is a very large number. Thus computation shows that

$$\begin{aligned} 2M(I_{27}) &= 0.3048\ldots \\ 2M(I_{28}) &= 208.06\ldots \\ 2M(I_{29}) &= 297{,}184.9\ldots \\ 2M(I_{30}) &= 904{,}093{,}985.9\ldots. \end{aligned}$$

Hence there are at least 209 distinct isomorphism classes in I_{28}, at least 297185 in I_{29}, and so on. The number $M(I_n)$ is asymptotic to

$$C(n/2\pi e\sqrt{e})^{n^2/4}(8\pi e/n)^{n/4}/\sqrt[24]{n}$$

as $n \to \infty$, where the constant C is approximately 0.705, and $2\pi e\sqrt{e} = 28.159\ldots$.

The actual details of the computation of $M(I_n)$ are rather tedious. We mention merely that the factors $\frac{1}{2}Df_p^{-1}(I)$ coming from finite primes are reasonable close to 1, and contribute very little to the final result. The manic behavior of the function $n \mapsto M(I_n)$ is entirely due to the single factor $\frac{1}{2}Df_\infty^{-1}(I)$. This factor can be computed by the formula

$$Df_\infty^{-1}(I) = \frac{1}{2}\omega_1 \frac{2}{2}\omega_2 \frac{3}{2}\omega_3 \ldots \frac{n}{2}\omega_n.$$

Further details will be omitted.

Similar computations can be carried out for spaces of type II. Compare [Serre, p. 94].

To conclude this section we must prove Lemmas (9.1) and (9.4).

Let $N_n(a)$ denote the number of solutions to the congruence

$$x_1^2+\cdots+x_n^2\equiv a \pmod p,$$

where p is a fixed odd prime. In [Siegel, p. 344] such numbers are computed by an ingenious Gauss sum argument. We will use an alternative method.

Let (a/p) be the Legendre symbol, equal to $+1$ or -1 according as a is or is not a quadratic residue modulo p. It will be convenient to extend the usual definition by agreeing that $(a/p)=0$ whenever $a\equiv 0 \pmod p$. With this convention, the number $N_1(a)$ of solutions to the congruence

$$x^2\equiv a \pmod p$$

is evidently equal to $1+(a/p)$.

To compute $N_2(a)$ we use the identity

$$N_2(a)=\sum_{x+y\equiv a \pmod p} N_1(x)\,N_1(y).$$

Substituting $1+(x/p)$ for $N_1(x)$ this yields

$$N_2(a)=\sum_{x+y\equiv a}\left(1+\left(\frac{x}{p}\right)+\left(\frac{y}{p}\right)+\left(\frac{xy}{p}\right)\right).$$

But clearly the sum of (x/p) over all residue classes x modulo p is equal to zero. Hence:

$$N_2(a)=p+0+0+\sum_{x+y\equiv a}\left(\frac{xy}{p}\right).$$

First suppose that $a=0$. Then each summand $\left(\frac{xy}{p}\right)=\left(\frac{-x^2}{p}\right)$ is equal to $\left(\frac{-1}{p}\right)$ if $x\not\equiv 0 \pmod p$, and is zero if $x\equiv 0$. Therefore

$$N_2(0)=p+(p-1)\left(\frac{-1}{p}\right). \tag{1}$$

On the other hand if a is relatively prime to p then the sum

$$\sum_{x+y\equiv a}\left(\frac{xy}{p}\right)$$

is independent of a. For if $x+y\equiv a$ then $ux+uy\equiv ua$ and

$$\left(\frac{ux\,uy}{p}\right)=\left(\frac{xy}{p}\right)$$

for any u relatively prime to p. Thus

$$N_2(1)=N_2(2)=\cdots=N_2(p-1).$$

But the sum

$$N_2(0)+N_2(1)+\cdots+N_2(p-1)$$

is clearly equal to p^2, the total number of pairs (x_1, x_2) mod p. Therefore

$$N_2(u)=N_2(1)=(p^2-N_2(0))/(p-1)$$

for any relatively prime residue class u modulo p. Substituting (1), we obtain the formula

$$N_2(u)=p-\left(\frac{-1}{p}\right). \tag{2}$$

It will be convenient to introduce the abbreviation $s=\left(\frac{-1}{p}\right)p^{-1}$, and to write these two formulas as

$$N_2(u)=p(1-s)$$

and

$$N_2(0)=N_2(u)+p^2 s.$$

Now using the identity

$$N_{2m+2}(a)=\sum_{x+y\equiv a} N_{2m}(x)\,N_2(y)$$

a straightforward induction on m shows that

$$N_{2m}(u)=p^{2m-1}(1-s^m) \tag{3}$$

and

$$N_{2m}(0)=N_{2m}(u)+p^{2m}s^m \tag{4}$$

for every positive integer m.

More generally, let $N_n(a \bmod p^k)$ denote the number of n-tuples $x_1, \ldots, x_n$ modulo p^k satisfying the congruence

$$x_1^2+\cdots+x_n^2\equiv a \pmod{p^k}. \tag{5}$$

These numbers can be computed as follows. We will assume that $a=p^v u$ with $v<k$, where u is relatively prime to p.

Case 1. If $v=0$, so that $a=u$ is a p-adic unit, then the number $N_n(u \bmod p^k)$ is equal to

$$p^{(n-1)(k-1)}N_n(u).$$

For each solution to the congruence

$$\xi_1^2+\cdots+\xi_n^2\equiv u \pmod p$$

gives rise to precisely $p^{(n-1)(k-1)}$ solutions modulo p^k. To see this, note that the ξ_i cannot all be divisible by p. If say $\xi_1 \not\equiv 0 \pmod p$, then choosing any residue classes $x_2, \ldots, x_n$ modulo p^k with

$$x_2\equiv\xi_2, \ldots, x_n\equiv\xi_n \pmod p,$$

it follows easily that the residue class of x_1 modulo p^k is uniquely determined.

Case 2. Even if $v\geqq 1$ there may still be solutions to the congruence

$$\xi_1^2+\cdots+\xi_n^2\equiv p^v u\equiv 0 \pmod p$$

in which the ξ_i are not all divisible by p. The total number of such solutions is equal to $N_n(0)-1$. Proceeding as in Case 1, the number of solutions to (5) in which the x_i are not all divisible by p is equal to $p^{(n-1)(k-1)}(N_n(0)-1)$.

Case 3. If $v\geqq 2$, then the Eq. (5) may have solutions in which all of the x_i are divisible by p. The total number of such solutions is equal to

$$p^n N_n(p^{v-2}u \bmod p^{k-2}),$$

since every solution to the congruence

$$\xi_1^2+\cdots+\xi_n^2\equiv p^{v-2}u \pmod{p^{k-2}}$$

gives rise to precisely p^n solutions to (5) satisfying the conditions

$$x_1\equiv p\xi_1, \ldots, x_n\equiv p\xi_n \pmod{p^{k-1}}.$$

Combining Cases 1, 2, 3 with the explicit formulas (3) and (4) for $N_{2m}(u)$ and $N_{2m}(0)$, a straightforward induction on v now produces the following conclusion.

(9.8) Lemma. *If* $0\leqq v<k$, *then the number* $N_{2m}(p^v u \bmod p^k)$ *of solutions to the congruence*

$$x_1^2+\cdots+x_{2m}^2\equiv p^v u \pmod{p^k}$$

is equal to

$$p^{(2m-1)k}(1-s^m)(1+p\,s^m+p^2 s^{2m}+\cdots+p^v s^{vm}),$$

where $s=\left(\frac{-1}{p}\right)p^{-1}$.

Proof of Lemma (9.1) *for p odd.* Let $Df_p^{-1}\colon \mathbf{Z}_p \to \mathbf{R}$ be the density function associated with $f_p(x_1, \ldots, x_{2m}) = x_1^2 + \cdots + x_{2m}^2$. Comparing the definition of Df_p^{-1} with (9.8), it follows easily that

$$Df_p^{-1}(p^v u) = (1-s^m)(1+p\,s^m + p^2 s^{2m} + \cdots + p^v s^{vm}).$$

Inspection shows that this equals the expression used in the statement of (9.1).

The proof for $p=2$ is similar. First note that the number $N_n(a \bmod 8)$ of solutions to the congruence $x_1^2 + \cdots + x_n^2 \equiv a \pmod 8$ is equal to

$$2^{3n-3} + 2^{5n/2-2} \cos\left(\frac{\pi}{4}(2a-n)\right)$$

plus a correction term of

$$(-1)^{(a-n)/4}\, 2^{2n-1} \qquad \text{if } a \equiv n \pmod 4.$$

In fact, the number of solutions to this congruence for which precisely t of the numbers $x_1, \ldots, x_n$ are odd is easily seen to be:

$$\binom{n}{t} 2^{2n-1} \qquad \text{if } 0 \leqq t < n \quad \text{and} \quad t \equiv a \pmod 4,$$

$$2^{2n} \qquad \text{if } t = n \equiv a \pmod 8,$$

and

$$0 \qquad \text{if } t \not\equiv a \pmod 4 \quad \text{or} \quad t = n \not\equiv a \pmod 8.$$

The sum, over all $t \equiv a \pmod 4$, of the binomial coefficient $\binom{n}{t}$ can be evaluated by forming the binomial expansions of $(1 \pm 1)^n$ and of $(1+i)^n$. The indicated expression for $N_n(a \bmod 8)$ follows easily.

Just as in the case p odd, one can now compute the number

$$N_n(a \bmod 2^k)$$

for large k, and hence compute the density of 2-adic solutions. Details will be left to the reader. ∎

Proof of (9.4). It will be convenient to emphasize dependence on n by using the symbol $\delta_n = \delta_{n,p}$ for the density function $Df_p^{-1}\colon \mathbf{Z}_p \to \mathbf{R}$ associated with the sum of n squares. We must find upper and lower bounds for $\prod_p \delta_{n,p}(k)$. First consider the case $n=8$, so that

$$\delta_8(p^v u) = (1-p^{-4})(1+p^{-3}+p^{-6}+\cdots+p^{-3v})$$

for p odd. Evidently δ_8 attains its minimum $1-p^{-4}$ when $v=0$, and tends to its maximum

$$(1-p^{-4})(1+p^{-3}+p^{-6}+\cdots) = (1-p^{-4})/(1-p^{-3})$$

as $v \to \infty$ (or as $p^v u \to 0$). Similarly, for $p=2$ the function

$$\delta_{8,2}(2^v u) = 1 + 8^{-1} + 8^{-2} + \cdots + 8^{1-v} - 8^{-v}$$

attains its minimum $\frac{7}{8}$ when $v=1$, and tends to its maximum $\frac{8}{7}$ as $v \to \infty$. Hence the product

$$\prod_{p \text{ finite}} \delta_{8,p}(k),$$

where k varies over $\mathbf{Z}$, attains its minimum

$$\tfrac{7}{8} \prod_{p \text{ odd}} (1 - p^{-4}) = \tfrac{14}{15} \zeta(4)^{-1} = 0.862\ldots$$

at $k=2$, and its maximum

$$\tfrac{8}{7} \prod_{p \text{ odd}} (1 - p^{-4})/(1 - p^{-3}) = \tfrac{16}{15} \zeta(3)/\zeta(4) = 1.184\ldots$$

at $k=0$. Since these bounds lie between $\frac{5}{6}$ and $\frac{6}{5}$, this completes the proof of (9.4) for the case $n=8$.

But any upper or lower bound for $\delta_{8,p}$ is automatically an upper or lower bound for $\delta_{n,p}$ for any $n \geqq 8$. In fact, dividing both sides of the identity

$$N_{m+n}(a \bmod p^k) = \sum_{x+y \equiv a (\mathrm{mod}\, p^k)} N_m(x \bmod p^k)\, N_n(y \bmod p^k)$$

by $p^{(m+n-1)k}$ and passing to the limit as $k \to \infty$, we obtain the convolution formula

$$\delta_{m+n}(a) = \int \delta_m(x)\, \delta_n(a-x)\, dx.$$

Thus, if $\delta_8(x) \leqq c$ for all x, it follows immediately that

$$\delta_{m+8}(a) \leqq \int \delta_m(x)\, c\, dx = c$$

for all a. This completes the proof. □

Chapter III. Inner Product Spaces over a Field

This chapter will describe some of the highlights of the theory of the Witt ring $W(F)$, where F is an arbitrary field. We are particularly careful to give proofs which are valid also in characteristic 2. The classical theory for number fields, as described for example in [O'Meara], is largely ignored.

All inner product spaces are to be symmetric.

§ 1. Anisotropic Inner Product Spaces

An inner product space X is *anisotropic* if $x \cdot x = 0$ implies $x = 0$. In this section we will show that every element in the Witt ring $W(F)$ is represented by an anisotropic inner product space which is unique up to isomorphism. First note the following.

(1.1) Lemma. *Every inner product space X over F is isomorphic to an orthogonal sum $S \oplus A$ with S split and A anisotropic.*

Proof. If the space X itself is anisotropic, then we can simply choose $S = 0$. If not, then there exists a vector $x \neq 0$ with $x \cdot x = 0$, and we can choose a vector y with $x \cdot y = 1$. Then x and y span a subspace $S_1 \subset X$ with inner product matrix $\begin{pmatrix} 0 & 1 \\ 1 & * \end{pmatrix}$. Evidently S_1 is split, and

$$X \cong S_1 \oplus S_1^{\perp}.$$

Applying the same construction to $S_1^{\perp}$, and continuing inductively, one can easily complete the proof. ▯

Choosing such a decomposition $X \cong S \oplus A$, we want to show that the anisotropic summand A is unique up to isomorphism. If the characteristic of F is $\neq 2$, then this is quite easy to prove, using Chapter I, Theorem (4.4) and Lemma (6.3). We will give a different argument which works also in characteristic 2.

Define the *index of isotropy* $i(X)$ of an inner product space X to be the maximal dimension of a subspace $N \subset X$ which is self-orthogonal, that is, $N \cdot N = 0$.

(1.2) Lemma. *The index of isotropy, for any inner product space X, satisfies*

$$0 \leqq 2i(X) \leqq \mathrm{rk}(X)$$

where the first equality holds if and only if X is anisotropic, and the second holds if and only if X is split.

Proof. If $N \cdot N = 0$, then $N \subset N^{\perp}$, and therefore

$$\mathrm{rk}(N) \leqq \mathrm{rk}(N^{\perp}) = \mathrm{rk}(X) - \mathrm{rk}(N),$$

where equality holds if and only if $N = N^{\perp}$. The argument is now straightforward. ∎

We will need the following upper bound.

(1.3) Main lemma. *If A is anisotropic, and X is an arbitrary inner product space, then*

$$i(X \oplus A) + i(X) \leqq \mathrm{rk}(X).$$

The proof will be based on the following interpretation of $i(X \oplus A)$. We will call a linear mapping

$$f\colon X \to Y$$

an *anti-isometry* if $f(x) \cdot f(x') = -x \cdot x'$ for all x and x' in X.

(1.4) Lemma. *With X and A as above, there exists a self-orthogonal subspace $N \subset X \oplus A$ of rank n if and only if there exists a subspace $M \subset X$ of rank n and an anti-isometry $f\colon M \to A$.*

Proof. Given such an anti-isometry f, let $N \subset X \oplus \mathrm{A}$ denote the graph of f, consisting of all pairs (x, a) with $x \in M$ and $f(x) = a$. Given any two pairs $(x, f(x))$ and $(x', f(x'))$ in the graph, we have

$$(x, f(x)) \cdot (x', f(x')) = x \cdot x' + f(x) \cdot f(x') = 0.$$

Thus N is self-orthogonal of rank n, as required.

Conversely, given $N \subset X \oplus A$ with $N \cdot N = 0$, note that N intersects the anisotropic subspace $0 \oplus \mathrm{A}$ in the zero vector only. If the pairs (x, a) and (x, a') both belong to N, it follows that $a = a'$. Thus N can be identified with the graph of a linear mapping $f\colon M \to A$ for some submodule $M \subset X$; and since N is self-orthogonal it follows that f is an anti-isometry. ∎

Proof of (1.3); by induction on the rank of $X \oplus A$. We may assume that $A \neq 0$, since if $A=0$ the inequality follows from (1.2).

Given an anti-isometry $f\colon M \to A$ with $M \subset X$, we must find an upper bound for the rank of M. Let M_0 denote the kernel of f, and let M_1 denote a complementary direct summand, so that

$$M = M_0 + M_1, \qquad 0 = M_0 \cap M_1.$$

Then M_1 maps injectively into A, so M_1 is anisotropic. Therefore the inner product, restricted to M_1, is again an inner product; and

$$X \cong M_1^{\perp} \oplus M_1.$$

Applying the induction hypothesis, we conclude that

$$i(X) + i(M_1^{\perp}) \leqq \operatorname{rk}(M_1^{\perp}).$$

But $M_0 \cdot M = 0$, so M_0 is a self-orthogonal subspace of $M_1^{\perp}$, and

$$\operatorname{rk}(M_0) \leqq i(M_1^{\perp}).$$

Adding these two inequalities, and adding $\operatorname{rk}(M_1) - i(M_1^{\perp})$ to both sides, we obtain

$$i(X) + \operatorname{rk}(M) \leqq \operatorname{rk}(X).$$

Therefore, by (1.4),

$$i(X) + i(X \oplus A) \leqq \operatorname{rk}(X),$$

which completes the proof. ▯

As an example, suppose that $X = S$ is split. Then the inequality $i(S \oplus A) \leqq \operatorname{rk}(S) - i(S) = i(S)$ implies that $i(S \oplus \mathrm{A}) = i(S)$.

(1.5) Corollary. *Suppose that an orthogonal sum $S \oplus A$, with S split and A anisotropic, is itself split. Then $A = 0$.*

For substituting $i(S \oplus A) = \frac{1}{2}\operatorname{rk}(S \oplus A)$ into the inequality $i(S \oplus A) + i(S) \leqq \operatorname{rk}(S)$, we obtain $\frac{1}{2}\operatorname{rk}(S \oplus A) \leqq \operatorname{rk}(S) - i(S) = \frac{1}{2}\operatorname{rk}(S)$, hence $\operatorname{rk}(A) \leqq 0$. ▯

(1.6) Corollary. *An inner product space X represents the zero element of the Witt ring if and only if X is split.*

For if X belongs to the Witt class of 0, then there exist split spaces S' and S'' so that

$$X \oplus S' \cong 0 \oplus S''.$$

Setting

$$X \cong A \oplus S$$

by (1.1), it follows that $A \oplus S \oplus S'$ is split. Therefore $A = 0$, and X is split. ▯

(1.7) Theorem. *Every element of the Witt ring $W(F)$ is represented by one, and up to isomorphism only one, anisotropic inner product space.*

Proof. If two anisotropic inner product spaces A and A' belong to the same Witt class, $A \sim A'$, then we must prove that $A \cong A'$. Consider the inner product space $B' = \langle -1 \rangle \otimes A'$. Then $A' \oplus B'$ is split, so it follows from (1.6) that the space $A \oplus B' \sim A' \oplus B'$ is split. Therefore, according to (1.4), there exists a subspace $M \subset A$ and an anti-isometry

$$f\colon M \to B',$$

where

$$\operatorname{rk}(M) = i(A \oplus B') = \tfrac{1}{2} \operatorname{rk}(A \oplus B').$$

But M is anisotropic, so f has trivial kernel, and

$$\operatorname{rk}(M) \leqq \operatorname{rk}(A), \qquad \operatorname{rk}(M) \leqq \operatorname{rk}(B').$$

It follows that $\operatorname{rk}(M) = \operatorname{rk}(A) = \operatorname{rk}(B')$, hence $M = A$, and f is an anti-isometry from A onto B'. Thus A is anti-isometric to B', and therefore A is isomorphic to A', which completes the proof. □

Thus the Witt ring $W(F)$ can be identified with the collection of isomorphism classes of anisotropic inner product spaces over F. To complete the picture, one more element of structure is needed.

(1.8) Definition. For each element w in the Witt ring $W(F)$, let $\|w\|$ denote the rank of the unique anisotropic representative for w.

Alternatively, choosing an arbitrary representative X for the Witt class w, we can set

$$\|w\| = \operatorname{rk}(X) - 2i(X).$$

Evidently

$$\|w\| \geqq 0$$

where equality holds only if $w = 0$. Note also the inequalities

$$\|w \pm w'\| \leqq \|w\| + \|w'\|,$$

$$\|w w'\| \leqq \|w\| \, \|w'\|$$

and

$$\|1\| = 1.$$

§ 2. Ordered Fields

This section is mainly a review of classical results concerning ordered fields. It studies the "total signature" $\sigma(X)$ of an inner product space over a field F. This is a certain Ω-tuple of integers, Ω being the collection of all orderings of F.

(2.1) Definition. An *ordering* of a field F is a subset $P \subset F^\bullet$ which is closed under addition and multiplication, and satisfies

$$P \cup (-P) = F^\bullet.$$

The elements of P are called *positive* (or *strictly positive*), and one writes $\xi > \eta$ if $\xi - \eta \in P$.

Note that the two subsets P and $-P$ are necessarily disjoint. For if both ξ and $-\xi$ belonged to P, then the sum $\xi + (-\xi)$ would have to belong to P, contradicting the hypothesis that $P \subset F^{\bullet}$.

In an ordered field, every non-zero square is positive. For if $\xi \neq 0$ then either $\xi \in P$ or $-\xi \in P$, and in either case it follows that $\xi^2 = (-\xi)^2 \in P$.

An ordered field necessarily has characteristic zero. For $1 = 1^2 \in P$, hence every sum $1 + 1 + \cdots + 1$ belongs to P. Therefore no such sum can equal 0 in F.

(2.2) Artin-Schreier theorem. *A field F possesses an ordering if and only if -1 is not a sum of squares in F.*

A field in which -1 is not a sum of squares is often called a "real" field, or a "formally real" field.

Proof. If F possesses an ordering P, then $1 \in P$, hence $-1 \notin P$, and -1 cannot be a sum of squares.

Conversely let F be a field in which -1 is not a sum of squares. By a *partial ordering* of F we will mean any subset of $F^{\bullet}$ which is closed under addition and multiplication. One partial ordering P_0 can be constructed as follows. Let P_0 be the collection of all sums of non-zero squares in F. Then P_0 is clearly closed under addition and multiplication, and if 0 belonged to P_0 then the equation

$$0 = \xi_1^2 + \cdots + \xi_n^2$$

with $\xi_i \neq 0$ would imply that

$$-1 = (\xi_2/\xi_1)^2 + \cdots + (\xi_n/\xi_1)^2,$$

contradicting our hypothesis.

By Zorn's lemma, the partial ordering P_0 is contained in some partial ordering P of F which is maximal with respect to inclusion. *Given $\xi \neq 0$, we will prove that either $\xi \in P$ or $-\xi \in P$.* This will show that P is an ordering of F. Consider the subset

$$Q = P \cup \xi P \cup (P + \xi P)$$

of F which is additively generated by P and ξP. If Q contains 0, then setting $0 = \pi' + \xi\pi$ with π' and π in P we obtain

$$-\xi = \pi'/\pi = \pi'\,\pi(\pi^{-1})^2 \in P.$$

(Every non-zero square belongs to P since $P \supset P_0$.) On the other hand, if Q does not contain 0 then evidently Q is a partial ordering of F. But Q contains the maximal partial ordering P, so $Q = P$, and it follows that $\xi \in P$. This shows that P is an ordering of F; which completes the proof. □

Here is an excursion for the reader.

(2.3) Exercise. Let F be a field of characteristic $\neq 2$. Using similar methods, prove that a field element $\xi \neq 0$ can be expressed as a sum of squares if and only if $\xi \in P$ for every ordering P of F. (Such element ξ are called "totally positive".) In particular, if F has no orderings at all, then every element of F is a sum of squares.

The classical case can be described as follows.

(2.4) Example. *If F is an algebraic extension of the rational numbers $\mathbf{Q}$, then every ordering of F is induced by an embedding $\varphi: F \to \mathbf{R}$. If F has finite degree n over $\mathbf{Q}$, then F has at most n distinct orderings.*

On the other hand for a field such as $\mathbf{Q}(\sqrt{2}, \sqrt{3}, \sqrt{5}, \dots)$ of infinite degree over $\mathbf{Q}$, there may well exist uncountably many distinct orderings.

Proof of (2.4). The rational numbers themselves have only one ordering, since every positive rational m/n can be expressed as an mn-fold sum of squares $n^{-2} + \dots + n^{-2}$. Thus every ordering of F induces the usual ordering in the subfield $\mathbf{Q} \subset F$. Note that no element ξ of F can satisfy $\xi > q$ for all $q \in \mathbf{Q}$. For if such an element ξ existed, then an easy induction would show that

$$\xi^n + q_1 \xi^{n-1} + \dots + q_n > 1$$

for every $n \geqq 1$, and every $q_1, \dots, q_n \in \mathbf{Q}$. [The inductive step consists of multiplying this inequality by ξ and noting that $\xi > 1 - q_{n+1}$.] Hence ξ could not be algebraic, contradicting our hypothesis.

Now every element ξ of F determines a Dedekind cut in $\mathbf{Q}$. Setting

$$\varphi(\xi) = \sup\{q \in \mathbf{Q} \mid q < \xi\} = \inf\{q \in \mathbf{Q} \mid q > \xi\},$$

it is clear that φ is additive; and that $\varphi(\xi\eta) = \varphi(\xi)\,\varphi(\eta)$, at least when $\xi > 0$ and $\eta > 0$. But the identities $\varphi(-\xi) = -\varphi(\xi)$ and $\varphi(1) = 1$ are also clear. Therefore $\varphi: F \to \mathbf{R}$ is a ring homomorphism, and φ has kernel zero since F is a field.

If F is finite over $\mathbf{Q}$, of degree n, then $F = \mathbf{Q}(\xi)$ for some element ξ, and the irreducible equation for ξ over $\mathbf{Q}$ can have at most n real roots. This completes the proof of (2.4). □

Now let us look at inner product spaces over a field F in which some ordering P has been specified. (Compare p. 23.)

Definition. An inner product space X over F is *positive definite* if $x \cdot x \in P$ for every non-zero vector x in X, and *negative definite* if $x \cdot x \in -P$ for every non-zero vector x in X.

(2.5) Inertia theorem (Jacobi, Sylvester). *Any inner product space X over an ordered field is isomorphic to an orthogonal sum $X^+ \oplus X^-$ with X^+*

positive definite and X^- negative definite. The ranks of X^+ and X^- are isomorphism invariants of X.

That is, these ranks do not depend on the particular choice of X^+ and X^-.

Proof. Choosing an orthogonal basis $e_1, \ldots, e_r$ for X, let X^+ be the subspace spanned by those e_i for which $e_i \cdot e_i > 0$, and let X^- be spanned by those e_i with $e_i \cdot e_i < 0$. Then evidently $X \cong X^+ \oplus X^-$ with X^+ positive definite and X^- negative definite.

Now let Y be an arbitrary positive definite subspace of X. Then $Y \cap X^- = 0$, hence

$$\operatorname{rk}(Y) \leqq \operatorname{rk}(X) - \operatorname{rk}(X^-) = \operatorname{rk}(X^+).$$

Therefore $\operatorname{rk}(X^+)$ can be characterized as the maximum possible dimension of a positive definite subspace of X. This shows that it is an isomorphism invariant of X, and completes the proof. ☐

Definition. The difference $\operatorname{rk}(X^+) - \operatorname{rk}(X^-)$ is called the *signature* of the inner product space X at the ordering P. We will use the notation

$$\sigma_P(X) \in \mathbf{Z}$$

for this signature. Evidently $\sigma_P(X)$ is also an isomorphism invariant of X.

In the case of an inner product space $\langle \alpha \rangle$ of rank 1, note that the signature $\sigma_P\langle \alpha \rangle$ is just what is usually called the *sign* of the field element α at the ordering P. That is $\sigma_P\langle \alpha \rangle$ is equal to $+1$ or -1 according as α is positive or negative.

(2.6) Lemma. *The signature $\sigma_P(X)$ depends only on the Witt class of X. Furthermore*

$$\sigma_P(X \oplus Y) = \sigma_P(X) + \sigma_P(Y),$$

$$\sigma_P(X \otimes Y) = \sigma_P(X)\, \sigma_P(Y),$$

and

$$\sigma_P\langle 1 \rangle = 1.$$

Thus the signature σ_P gives rise to a well defined homomorphism from the Witt ring $W(F)$ to the ring of integers $\mathbf{Z}$.

Proof. If S is a split inner product space, then it follows from Chapter I, § 6.3 that S is isomorphic to an orthogonal sum of copies of $\langle 1 \rangle \oplus \langle -1 \rangle$. Hence the signature $\sigma_P(S)$ is zero. [Or more directly, if $N \subset S$ is self-orthogonal then the argument used to prove (2.5) shows also that $\operatorname{rk}(N) \leqq \operatorname{rk}(S^+)$ and $\operatorname{rk}(N) \leqq \operatorname{rk}(S^-)$. So if $\operatorname{rk}(N) = \frac{1}{2}\operatorname{rk}(S)$, it follows that $\operatorname{rk}(N) = \operatorname{rk}(S^+) = \operatorname{rk}(S^-)$, and the signature is zero.]

Now expressing X and Y as sums of spaces of rank 1, and using the evident identity

$$\sigma_P(\langle\alpha_1\rangle\oplus\cdots\oplus\langle\alpha_r\rangle)=\sigma_P\langle\alpha_1\rangle+\cdots+\sigma_P\langle\alpha_r\rangle,$$

the rest of the proof is straightforward. □

(2.7) Corollary. *Suppose that F is an ordered field in which every positive element is a square. Then σ_P: $W(F)\to\mathbf{Z}$ is an isomorphism.*

For example the Witt ring of the real numbers $\mathbf{R}$ is isomorphic to $\mathbf{Z}$. The proof is straightforward from (2.5) and (2.6). □

Remark. It is shown in standard algebra texts (Lang or van der Waerden) that there is a one-to-one correspondence between orderings of a field F and isomorphism classes of "real closures" of F. The real closure F_P associated with any ordering P can be characterized as the maximal compatibly ordered algebraic extension field of F. Every positive element of F_P is a square, hence the Witt ring $W(F_P)$ is isomorphic to $\mathbf{Z}$. It follows easily that the signature homomorphism σ_P: $W(F)\to\mathbf{Z}$ can be identified with the natural ring homomorphism $W(F)\to W(F_P)$.

Now consider the collection $\Omega=\Omega(F)$ consisting of all possible orderings of a given field F. We topologize Ω as follows.

Definition. For each $\xi\in F^\bullet$ let $U_\xi\subset\Omega$ be the set of all orderings P for which $\xi\in$ P. Then these sets U_ξ generate the required topology. (In other words a subset of Ω is defined to be open if and only if it is a union of finite intersections of the U_ξ.)

(2.8) Lemma. *This topological space Ω is compact and totally disconnected. For each inner product space X over F, the function*

$$P\mapsto\sigma_P(X)$$

from Ω to $\mathbf{Z}$ is continuous.

Definition. This function $P\mapsto\sigma_P(X)$ will be called the *total signature* $\sigma(X)$ of the inner product space X.

Proof of (2.8). First consider an inner product space $\langle\xi\rangle$ of rank 1. Then the inverse image of 1 under the total signature function

$$P\mapsto\sigma_P\langle\xi\rangle$$

is the open set $U_\xi\subset\Omega$, and the inverse image of -1 is the complementary open set $U_{-\xi}$. Thus the total signature function of $\langle\xi\rangle$ is continuous, and it follows easily that the total signature function associated with any inner product space $X\cong\langle\xi_1\rangle\oplus\cdots\oplus\langle\xi_r\rangle$ is continuous.

Since Ω is the union of disjoint open sets U_ξ and $U_{-\xi}$ for any $\xi \in F^\bullet$, it is easy to check that Ω is Hausdorff and totally disconnected. To prove compactness, we introduce the space 2^F consisting of all subsets of F. This is to be topologized as a cartesian product. That is we identify each subset of F with its characteristic function $F \to \{0, 1\}$, and hence identify 2^F with a cartesian product of copies of $\{0, 1\}$, one copy for each element of F. This product is compact by Tychonoff's theorem.

Each ordering of F can be considered as an element of 2^F, so Ω is embedded as a subset of 2^F. Evidently the topology which we constructed for Ω is precisely the relative topology which Ω acquires as a subset of 2^F. In fact Ω is a closed subset of 2^F. For if a subset $Q \subset F$ is not an ordering, then it is easy to construct a neighborhood of Q in 2^F which does not contain any orderings. This proves that Ω is compact, and completes the proof of (2.8). □

(2.9) Definition. The ring consisting of all continuous functions from Ω to $\mathbf{Z}$ will be denoted by $\mathbf{Z}^\Omega$. Evidently, for any inner product space X over F, the total signature function $\sigma(X)$ is an element of this ring $\mathbf{Z}^\Omega$. Since σ is additive, multiplicative, and since $\sigma(X)$ depends only on the Witt class of X, we obtain a well defined ring homomorphism

$$\sigma\colon\ W(F) \to \mathbf{Z}^\Omega.$$

[Of course if the set Ω is vacuous, then $\mathbf{Z}^\Omega$ is the zero ring, and this construction is not particularly interesting.]

Remark. If Ω has more than one element, then this homomorphism σ is not surjective. In fact the congruence

$$\sigma_P(X) \equiv \mathrm{rk}(X) \pmod 2$$

clearly holds for any inner product space X and any ordering P. Therefore the total signature

$$\sigma(X) \in \mathbf{Z}^\Omega$$

must be congruent to either 0 or 1 modulo the ideal $2\mathbf{Z}^\Omega$, according as the rank of X is even or odd.

In order to determine the precise image of $\sigma\colon W(F) \to \mathbf{Z}^\Omega$, one needs to know to what extent it is possible to prescribe the signs of a field element at the various orderings of F.

(2.10) Example. *If F is an algebraic extension of the rationals, then given a completely arbitrary open and closed subset $U \subset \Omega$, there exists a field element α which satisfies the condition*

$$\alpha \in P \Leftrightarrow P \in U.$$

Now the signature $\sigma(\langle 1\rangle+\langle\alpha\rangle)\in\mathbf{Z}^{\Omega}$ is twice the characteristic function of U, and by adding such characteristic functions, one sees that *every element of the ideal* $2\mathbf{Z}^{\Omega}$ *belongs to the image of* σ. (The corresponding statement for a completely arbitrary field would be false.)

Proof of (2.10). First suppose that F has finite degree over $\mathbf{Q}$. Then there exist only finitely many embeddings

$$\varphi_1, \ldots, \varphi_m \colon F \to \mathbf{R},$$

and it clearly suffices to construct field elements $\alpha_1, \ldots, \alpha_m$ so that $\varphi_i(\alpha_j)$ is positive if $i \neq j$ and negative if $i=j$. Suitable products of the α_i will then have arbitrarily prescribed signs.

Choose a field element ξ so that $F=\mathbf{Q}(\xi)$. Now choose a rational number $\varepsilon>0$ so that

$$2\varepsilon < |\varphi_i(\xi)-\varphi_j(\xi)|$$

for $i \neq j$, and choose rational numbers $q_1, \ldots, q_m$ so that

$$|\varphi_i(\xi)-q_i|<\varepsilon.$$

The differences

$$\alpha_i=(\xi-q_i)^2-\varepsilon^2$$

will then have the required property.

Now suppose that F has infinite degree over $\mathbf{Q}$. The given open and closed set $U\subset\Omega$ can be covered by finitely many basic open sets $U_{\xi_1}\cap\cdots\cap U_{\xi_k}$ which do not intersect the complement of U. Taking all of these field elements $\xi_1, \ldots, \xi_k$ for all of the basic open sets which are needed, we generate a certain subfield $F_0\subset F$ which is finite over $\mathbf{Q}$. Clearly the given set $U\subset\Omega(F)$ is equal to the inverse image of a suitable subset $U_0\subset\Omega(F_0)$ under the restriction morphism

$$\Omega(F)\to\Omega(F_0).$$

Choosing $\alpha\in F_0$ so as to be positive at the orderings in U_0 and negative at the orderings in the complement of U_0, this completes the proof. □

For further information see [Knebusch, Rosenberg, and Ware].

§ 3. Prime Ideals in the Witt Ring

In this section we study the structure of the Witt ring $W(F)$ for an arbitrary field F. The results are due to Pfister, but the simplified proofs are due to Lorenz and Leicht.

(3.1) Lemma. *The Witt ring* $W(F)$ *is additively generated by the elements* $\langle\alpha\rangle$, *where* α *varies over* $F^{\bullet}$.

Proof. This follows immediately from Chapter I, § 3. □

(3.2) Lemma. *If $\mathfrak{p}$ is an arbitrary prime ideal in $W(F)$, then for every $\alpha \in F^\bullet$ either*

$$\langle\alpha\rangle \equiv \langle 1\rangle \quad \bmod \mathfrak{p}$$

or

$$\langle\alpha\rangle \equiv \langle -1\rangle \quad \bmod \mathfrak{p}.$$

Hence the quotient ring $W(F)/\mathfrak{p}$ is isomorphic either to $\mathbf{Z}$, or to the field $\mathbf{F}_p = \mathbf{Z}/p\mathbf{Z}$ for some prime number p.

Proof. Since the element $\langle\alpha^2\rangle$ of $W(F)$ is equal to $\langle 1\rangle$, we have

$$(\langle\alpha\rangle - \langle 1\rangle)(\langle\alpha\rangle + \langle 1\rangle) = 0$$

in $W(F)$. Therefore, modulo any prime ideal $\mathfrak{p}$, the element $\langle\alpha\rangle$ is congruent to either $\langle 1\rangle$ or $-\langle 1\rangle$. It follows that the unique ring homomorphism

$$\mathbf{Z} \to W(F) \to W(F)/\mathfrak{p}$$

is surjective. Since the kernel must be a prime ideal of $\mathbf{Z}$, this completes the proof. ▯

First consider the case $p = 2$.

(3.3) Lemma. *For any field F, there is one and only one ideal I in $W(F)$ such that $W(F)/I \cong \mathbf{F}_2$.*

Thus $I = I(F)$ is the kernel of the unique ring homomorphism $W(F) \to \mathbf{F}_2$. We will call I the *fundamental ideal* of the Witt ring.

Proof of (3.3). If $W(F)/I \cong \mathbf{F}_2$, then

$$\langle 1\rangle \equiv \langle -1\rangle \quad \bmod I,$$

and therefore

$$\langle\alpha\rangle \equiv \langle 1\rangle \quad \bmod I$$

for every α. It follows that I consists precisely of all sums $\langle\alpha_1\rangle + \cdots + \langle\alpha_r\rangle$ for which the rank r is even.

Conversely, since every split inner product space has even rank, the correspondence

$$X \mapsto \operatorname{rk}(X) \quad \text{modulo } 2$$

gives rise to a ring homomorphism $W(F) \to \mathbf{F}_2$ whose kernel is the required fundamental ideal I. ▯

(3.4) Remark. *The Witt ring $W(F)$ is isomorphic to $\mathbf{F}_2$ if and only if every element of $F^\bullet$ is a square.*

Examples are provided by algebraically closed fields, and perfect fields of characteristic 2.

Proof. If $W(F) \cong \mathbf{F}_2$, then for every $\alpha \in F^\bullet$ the anisotropic inner product space $\langle \alpha \rangle$ belongs to the same Witt class as $\langle 1 \rangle$, and therefore is isomorphic to $\langle 1 \rangle$. Hence α is a square. Since the converse is straightforward, this completes the proof. □

Now let us look at the remaining prime ideals.

(3.5) Main lemma. *If* $\mathfrak{p} \subset W(F)$ *is any prime ideal with*

$$W(F)/\mathfrak{p} \not\cong \mathbf{F}_2,$$

then the set $P \subset F^\bullet$ *consisting of all field elements* $\alpha \neq 0$ *with*

$$\langle \alpha \rangle \equiv \langle 1 \rangle \pmod{\mathfrak{p}}$$

constitutes an ordering of F. *The associated signature homomorphism*

$$\sigma_P \colon W(F) \to \mathbf{Z}$$

satisfies the congruence

$$w \equiv \sigma_P(w) \langle 1 \rangle \pmod{\mathfrak{p}}$$

for every element w *in the Witt ring.*

Hence the given prime ideal $\mathfrak{p}$ is either the kernel of σ_P or the kernel of the composition

$$W(F) \xrightarrow{\sigma_P} \mathbf{Z} \to \mathbf{F}_p$$

according as the quotient $W(F)/\mathfrak{p}$ is isomorphic to $\mathbf{Z}$ or $\mathbf{F}_p$.

Proof. It is clear that P is closed under multiplication, and that

$$P \cup (-P) = F^\bullet.$$

So we need only prove that P is closed under addition. If $\alpha, \beta \in P$ and $\alpha + \beta = \gamma \neq 0$, then evidently

$$\langle \alpha \rangle \oplus \langle \beta \rangle \cong \langle \gamma \rangle \oplus \langle \delta \rangle$$

for some δ. Reducing modulo $\mathfrak{p}$ in the Witt ring, this yields

$$\langle 1 \rangle + \langle 1 \rangle \equiv \langle \gamma \rangle + \langle \delta \rangle \pmod{\mathfrak{p}}$$

with $\langle \gamma \rangle \equiv \langle \pm 1 \rangle$ and $\langle \delta \rangle \equiv \langle \pm 1 \rangle$. But the hypothesis that $W(F)/\mathfrak{p} \not\cong \mathbf{F}_2$ implies that $\langle 1 \rangle + \langle 1 \rangle$ is not congruent to either $\langle -1 \rangle + \langle 1 \rangle$ or $\langle -1 \rangle + \langle -1 \rangle$ modulo $\mathfrak{p}$. Therefore

$$\langle \gamma \rangle \equiv \langle 1 \rangle \pmod{\mathfrak{p}},$$

hence $\gamma = \alpha + \beta \in P$. The case $\alpha, \beta \in P$ and $\alpha + \beta = 0$ cannot occur, since the inner product space $\langle \alpha \rangle \oplus \langle \beta \rangle$ would then be split, implying that

$$\langle 1 \rangle + \langle 1 \rangle \equiv \langle \alpha \rangle + \langle \beta \rangle = 0$$

which contradicts our hypothesis. Thus P is an ordering of F.

Now for any α in $F^\bullet$ the congruence

$$\langle\alpha\rangle \equiv \sigma_P(\langle\alpha\rangle)\,\langle 1\rangle \qquad (\mathrm{mod}\ \mathfrak{p})$$

follows from the definition of P. It clearly follows that

$$w \equiv \sigma_P(w)\,\langle 1\rangle \qquad (\mathrm{mod}\ \mathfrak{p})$$

for every element $w = \langle\alpha_1\rangle + \cdots + \langle\alpha_r\rangle$ in the Witt ring. This completes the proof. ▯

Remark. It follows easily that the topological space $\mathrm{Spec}(W(F))$, consisting of all prime ideals of $W(F)$ with the Zariski topology, is homeomorphic to the quotient space which is obtained from the cartesian product $\Omega \times \mathrm{Spec}(\mathbf{Z})$ by collapsing the closed subset $\Omega \times (2)$ to a point. Each pair $(P, \mathfrak{q})$ in $\Omega \times \mathrm{Spec}(\mathbf{Z})$ corresponds of course to the kernel of $(\sigma_P \bmod \mathfrak{q})\colon\ W(F) \to \mathbf{Z}/\mathfrak{q}$.

One immediate consequence of 3.5 is the following.

(3.6) Theorem. *If -1 is a sum of squares in F, then $W(F)$ is a local ring containing just one prime ideal I. Every element of I is nilpotent; and the order of every element in the additive group of $W(F)$ is a power of 2.*

Before giving the proof, let us review some elementary ring theory.

(3.7) Lemma. *For any commutative ring R, the intersection of all prime ideals in R is precisely equal to the ideal $\mathfrak{N}$ consisting of all nilpotent elements in R.*

For if $\xi^n = 0$, then certainly ξ belongs to every prime ideal of R. Conversely suppose that ξ is not nilpotent. Consider all ideals $\mathfrak{a} \subset R$ with the property that no power of ξ lies in $\mathfrak{a}$. This collection of ideals is non-vacuous, since the zero ideal has this property. By Zorn's lemma, there exists such an ideal $\mathfrak{a}$ which is maximal with respect to inclusion. Then $\mathfrak{a}$ is a prime ideal. For if $\alpha \notin \mathfrak{a}$ and $\beta \notin \mathfrak{a}$ then the ideals $\mathfrak{a} + R\alpha$ and $\mathfrak{a} + R\beta$ are strictly larger than $\mathfrak{a}$. Therefore

$$\xi^m \in \mathfrak{a} + R\alpha, \qquad \xi^n \in \mathfrak{a} + R\beta$$

for some m and n. Hence

$$\xi^{m+n} \in (\mathfrak{a} + R\alpha)(\mathfrak{a} + R\beta) \subset \mathfrak{a} + R\alpha\beta.$$

But no power of ξ lies in $\mathfrak{a}$, so this proves that $\alpha\beta \notin \mathfrak{a}$. Thus $\mathfrak{a}$ is a prime ideal. Hence the given element $\xi \notin \mathfrak{a}$ does not lies in the intersection of all prime ideals. This completes the proof of (3.7). ▯

Proof of (3.6). Since F cannot be ordered, the Witt ring $W(F)$ cannot have any prime ideals $\mathfrak{p}$ other than the fundamental ideal I. Therefore the ideal $\mathfrak{N}$ of nilpotent elements, being the intersection of all prime ideals, is precisely equal to I.

In particular, since $\langle 1\rangle + \langle 1\rangle \in I$, it follows that some power $(\langle 1\rangle + \langle 1\rangle)^n = 2^n\langle 1\rangle$ is zero in $W(F)$. (Here $2^n\langle 1\rangle$ denotes the sum of 2^n copies of $\langle 1\rangle$.) Therefore $2^n w = 0$ for every w in the Witt ring. This completes the proof. ∎

Remark. This proof makes rather blatant use of Zorn's lemma. We will give a more constructive argument in § 4.6.

Now consider a field in which -1 is not a sum of squares. Let $\mathfrak{N}$ denote the ideal consisting of all nilpotent elements in the Witt ring.

(3.8) Theorem. *If -1 is not a sum of squares in F, then the nilradical $\mathfrak{N} \subset W(F)$ is precisely equal to the kernel of the total signature homomorphism*

$$\sigma:\ W(F) \to \mathbf{Z}^{\Omega}$$

of § 2.8. An element w in the Witt ring is a unit if and only if its image $\sigma(w)$ is a unit in the ring $\mathbf{Z}^{\Omega}$.

Proof. If $\sigma(w) = 0$, then certainly w belongs to the fundamental ideal I, consisting of Witt classes of even rank. But it follows from (3.5) that w belongs to every other prime ideal also. Therefore $w \in \mathfrak{N}$. Conversely, if $w \in \mathfrak{N}$ then $\sigma(w)$ is nilpotent hence $\sigma(w) = 0$.

Now suppose that $\sigma(w)$ is a unit. Then $\sigma(w)^2 = 1$, hence $w^2 \equiv 1 \bmod \mathfrak{N}$, and it follows that w is a unit. This completes the proof. ∎

(3.9) Corollary. *The Witt ring $W(F)$ is isomorphic to $\mathbf{Z}$ if and only if F is an ordered field in which every positive element is a square.*

For if $W(F) \cong \mathbf{Z}$, then F can be ordered, and for every $\alpha > 0$ the anisotropic inner product space $\langle\alpha\rangle$ belongs to the same Witt class as $\langle 1\rangle$ and hence is isomorphic to $\langle 1\rangle$. Therefore α is a square. Together with (2.7), this completes the proof. ∎

(3.10) Theorem. *For any field F, the torsion subgroup of $W(F)$ is precisely the kernel of the total signature homomorphism*

$$\sigma:\ W(F) \to \mathbf{Z}^{\Omega}.$$

The order of every torsion element is a power of 2.

Remark. If F has only one ordering, then this can be proved quite simply as follows. Clearly the kernel of $\sigma: W(F) \to \mathbf{Z}$ is additively generated by elements of the form $\langle 1\rangle - \langle\alpha\rangle$ with $\alpha > 0$. But

$$(\langle 1\rangle - \langle\alpha\rangle)^2 = 2(\langle 1\rangle - \langle\alpha\rangle)$$

and therefore

$$(\langle 1\rangle - \langle\alpha\rangle)^n = 2^{n-1}(\langle 1\rangle - \langle\alpha\rangle).$$

Since $\langle 1\rangle - \langle\alpha\rangle$ is known to be nilpotent, it follows that its order is a power of 2.

The proof in the general case will be based on the following. Let K be any extension field of F. Recall from Chapter I, § 5.4 that any inner product space X of rank r over F gives rise to an inner product space $K \otimes_F X$ of rank r over K. Clearly this correspondence induces a ring homomorphism

$$i_*: W(F) \to W(K).$$

For a quadratic extension, the kernel of this homomorphism is computed as follows. We assume that F has characteristic $\neq 2$.

(3.11) Lemma. *For any $\alpha \in F^\bullet$, the kernel of the natural homomorphism $W(F) \to W(F(\sqrt{\alpha}))$ is equal to the principal ideal $(\langle 1\rangle - \langle\alpha\rangle)\, W(F)$. Every element w in the kernel satisfies $w = -\langle\alpha\rangle\, w$.*

Proof. Certainly α maps to a square in $F(\sqrt{\alpha})$, hence the ideal $(\langle 1\rangle - \langle\alpha\rangle)\, W(F)$ maps to zero. But if an anisotropic inner product space $X \neq 0$ over F represents a Witt class in the kernel, then there certainly exists a vector $z \neq 0$ in $F(\sqrt{\alpha}) \otimes_F X$ with $z \cdot z = 0$. Setting $z = x + \sqrt{\alpha}\, y$ with x and y in X, the equation $z \cdot z = 0$ implies that

$$x \cdot x + \alpha\, y \cdot y = 0, \qquad x \cdot y = 0.$$

Since X is anisotropic, at least one of the field elements $x \cdot x$ and $y \cdot y$ is non-zero, hence both are non-zero. Setting $y \cdot y = \eta_1$ and $x \cdot x = -\alpha\, \eta_1$, we see that X decomposes as an orthogonal sum

$$\langle\eta_1\rangle \oplus \langle -\alpha\, \eta_1\rangle \oplus X',$$

where X' is also anisotropic and also represents an element of the kernel. An easy induction now shows that

$$X \cong (\langle 1\rangle \oplus \langle -\alpha\rangle) \otimes (\langle\eta_1\rangle \oplus \cdots \oplus \langle\eta_k\rangle)$$

for suitable $\eta_1, \ldots, \eta_k$. Clearly it follows that

$$X \cong \langle -\alpha\rangle \otimes X,$$

which completes the proof. □

Proof of Theorem (3.10). Suppose that some nilpotent element w of $W(F)$ satisfies $2^n w \neq 0$ for all n. Consider algebraic extension fields $K \supset F$ such that the image $w' = i_*(w)$ in $W(K)$ satisfies $2^n w' \neq 0$ for all n. Note that any monotone union of fields having this property will again have this property. Hence, by Zorn's lemma, there exists a maximal extension field K with this property. If $\alpha \in K$ is any non-square, then the field $K(\sqrt{\alpha})$ is strictly larger than K. Therefore the image of $2^n w'$ in $W(K(\sqrt{\alpha}))$ is zero for large n, and it follows from (3.11) that

$$2^n w' = \langle -\alpha\rangle\, 2^n w'$$

for large n.

Since $W(K)$ is not a 2-torsion group, it follows from (3.6) that K can be ordered. Since $W(K)$ possesses nilpotent elements, it follows from (3.9) that not every positive element of K is a square. Therefore the quotient $K^\bullet/K^{\bullet 2}$ contains at least four distinct elements. Let 1, α, β, and $\alpha\beta$ in $K^\bullet$ be distinct modulo $K^{\bullet 2}$. Then

$$\begin{aligned} 2^n w' &= \langle -\alpha \rangle 2^n w' = \langle -\alpha \rangle \langle -\beta \rangle 2^n w' \\ &= \langle -\alpha \rangle \langle -\beta \rangle \langle -\alpha\beta \rangle 2^n w' \end{aligned}$$

for large n. This proves that $2^n w' = -2^n w'$ or $2^{n+1} w' = 0$, which contradicts our hypothesis and completes the proof. ☐

(3.12) Corollary. *For each ordering P let F_P denote the associated real closure of F. Then the kernel of the natural homomorphism*

$$W(F) \to \prod_P W(F_P)$$

is equal to the torsion subgroup of $W(F)$.

The proof is immediate.

To conclude this section, we will outline a different description of the torsion in $W(F)$, due to Scharlau [1969], [1970]. Let F be a field of charactersistic $\neq 2$.

Definition. The field F is *pythagorean* if the subset F^2 is closed under addition; or in other words if every sum of squares is a square in F.

Given any field F of characteristic $\neq 2$, with algebraic closure $\bar{F}$, there is a unique smallest extension field $F_{py} \subset \bar{F}$ which is pythagorean. In fact F_{py} is the union of all iterated quadratic extensions of the form

$$F \subset \cdots \subset K \subset K\left(\sqrt{\alpha^2 + \beta^2}\right)$$

within $\bar{F}$. We will call this unique field F_{py} the *pythagorean closure* of F.

The Witt ring $W(F_{py})$ of a pythagorean field can be described as follows. If -1 is a sum of squares in F_{py}, then every element of F_{py} is a square, and it follows that $W(F_{py}) \cong \mathbf{Z}/2\mathbf{Z}$. (Compare (2.3) and (3.4).) *On the other hand if -1 is not a sum of squares, then $W(F_{py})$ is torsion free.* In fact, given $w \neq 0$ in $W(F_{py})$ *choose an anisotropic representative* $A \cong \langle \alpha_1 \rangle \oplus \cdots \oplus \langle \alpha_n \rangle$ for w. Then for any $k > 0$ the k-fold sum $A \oplus \cdots \oplus A$ is also anisotropic. For if the equation $\sum_{j=1}^{k} \sum_{i=1}^{n} \alpha_i \xi_{ij}^2 = 0$ had a non-trivial solution then, setting $\sum_j \xi_{ij}^2 = \eta_i^2$, it would follow that the equation $\sum \alpha_i \eta_i^2 = 0$ also had a non-trivial solution, which is impossible.

(3.13) Assertion. *If F_{py} is the pythagorean closure of F, then the sequence*

$$0 \to \operatorname{Tors} IW(F) \to W(F) \to W(F_{py})$$

is exact.

Here Tors $IW(F)$ denotes the torsion subgroup of the fundamental ideal $I \subset W(F)$, considered as additive group.

Proof. First consider a quadratic extension of the form

$$K \subset K\left(\sqrt{\alpha^2+\beta^2}\right).$$

By (3.11) the kernel of the associated homomorphism

$$W(K) \to W\left(K\left(\sqrt{\alpha^2+\beta^2}\right)\right)$$

is equal to the ideal $(\langle 1\rangle - \langle \alpha^2+\beta^2\rangle)\, W(K)$. Using the isomorphism

$$\langle 1\rangle \oplus \langle 1\rangle \cong \langle \alpha^2+\beta^2\rangle \oplus \langle \alpha^2+\beta^2\rangle$$

we see that every element in this ideal has order 2. For an iterated extension $F \subset \cdots \subset K \subset K\left(\sqrt{\alpha^2+\beta^2}\right)$ of degree 2^n over F it follows inductively that the kernel of the associated homomorphism of Witt rings has exponent 2^n. Passing to the direct limit (= inductive limit) of all such iterated extensions, it follows that the kernel of the homomorphism

$$W(F) \to W(F_{py})$$

is a 2-primary torsion group. But $W(F_{py})$ is torsion free, so this kernel must be precisely equal to the torsion subgroup of $W(F)$. □

This argument, incidentally, gives a simpler proof that the Witt ring of a field contains no odd torsion.

Note that for any field (even in characteristic 2) the subgroup Tors $IW(F)$ is precisely equal to the nilradical of $W(F)$.

§ 4. Multiplicative Inner Product Spaces

The results in this section are due to Pfister. (Compare [Scharlau, 1969] and [Lorenz].) However for convenience we will modify Pfister's definitions.

If x belongs to an inner product space X, it will be convenient to call $x \cdot x$ the *norm* of x. Thus a field element α is a *norm from* X if $\alpha = x \cdot x$ for some x.

(4.1) Definition. An inner product space X is *multiplicative* if

$$X \cong \langle \alpha\rangle \otimes X$$

for every field element $\alpha \neq 0$ which is a norm from X.

(This is not the usual definition.)

One important property of multiplicative spaces is the following.

(4.2) Lemma. *If X is multiplicative, then the set of all field elements $\alpha \neq 0$ which are norms from X forms a subgroup of $F^{\bullet}$.*

Proof. If $\alpha = x \cdot x \neq 0$ and $\beta = y \cdot y \neq 0$, choose an isomorphism

$$f\colon X \to \langle\beta\rangle \otimes X.$$

Setting $f(x) = e \otimes z$, where $e \cdot e = \beta$, we obtain

$$x \cdot x = f(x) \cdot f(x) = \beta\, z \cdot z.$$

Therefore the quotient $\alpha/\beta = z \cdot z$ is also a norm from X, which completes the proof. $\square$

As an example, the inner product space $\langle 1\rangle$ is certainly multiplicative, with group of non-zero norms equal to $F^{\bullet 2}$.

(4.3) Theorem. *Any tensor product of the form*

$$(\langle 1\rangle \oplus \langle\alpha_1\rangle) \otimes \cdots \otimes (\langle 1\rangle \oplus \langle\alpha_n\rangle)$$

is multiplicative. Furthermore any such tensor product is either anisotropic or split.

Proof. First consider the case $n=1$. If $\beta \neq 0$ is a norm from $\langle 1\rangle \oplus \langle\alpha\rangle$, then clearly

$$\langle 1\rangle \oplus \langle\alpha\rangle \cong \langle\beta\rangle \oplus \langle\gamma\rangle$$

for some γ. Comparing determinants, we see that $\langle\gamma\rangle \cong \langle\beta\,\alpha\rangle$, so

$$\langle 1\rangle \oplus \langle\alpha\rangle \cong \langle\beta\rangle \otimes (\langle 1\rangle \oplus \langle\alpha\rangle)$$

as required. Since very inner product space of rank 2 is either anisotropic or split, this takes care of the case $n=1$.

The proof now proceeds by induction. Assuming that X is multiplicative, we will show that the space

$$(\langle 1\rangle \oplus \langle\alpha\rangle)\, X \cong X \oplus \langle\alpha\rangle\, X$$

is multiplicative. Here we are leaving out the tensor product signs, to simplify the notation. Let $\beta \neq 0$ be a norm from $X \oplus \langle\alpha\rangle\, X$. Then clearly β has the form

$$\beta = x \cdot x + \alpha\, y \cdot y = \xi + \alpha\,\eta$$

where ξ and η are norms from X. If $\xi = 0$, then $\eta \neq 0$, so $\langle\eta\rangle\, X \cong X$ and $\langle\alpha\,\eta\rangle\,(X \oplus \langle\alpha\rangle\, X) \cong \langle\alpha\rangle\, X \oplus \langle\alpha\rangle^2\, X \cong X \oplus \langle\alpha\rangle\, X$, as required. The case $\eta = 0$ is handled similarly. Suppose then that both ξ and η are non-zero. Then $X \cong \langle\xi\rangle\, X \cong \langle\eta/\xi\rangle\, X$ so it follows that

$$\begin{aligned}\langle\xi + \alpha\,\eta\rangle\,(X \oplus \langle\alpha\rangle\, X) &\cong \langle 1 + \alpha\,\eta/\xi\rangle\,(X \oplus \langle\alpha\,\eta/\xi\rangle\, X)\\ &\cong \langle 1 + \alpha\,\eta/\xi\rangle\,(\langle 1\rangle \oplus \langle\alpha\,\eta/\xi\rangle)\, X.\end{aligned}$$

But we have already established that the inner product space

$$\langle 1\rangle \oplus \langle \alpha\eta/\xi\rangle$$

is multiplicative. Since $1+\alpha\eta/\xi$ is a non-zero norm from this space, we see that the factor $\langle 1+\alpha\eta/\xi\rangle$ can be cancelled, and we are left with

$$(\langle 1\rangle \oplus \langle \alpha\eta/\xi\rangle)\, X \cong X \oplus \langle \alpha\rangle X,$$

as required.

We must prove also that $X \oplus \langle \alpha\rangle X$ is either anistropic or split, assuming inductively that X itself is either anisotropic or split. If $X \oplus \alpha X$ is not anisotropic, then there exist vectors x and y in X, not both zero, so that

$$x\cdot x+\alpha\, y\cdot y=\xi+\alpha\eta=0.$$

If $\xi=\eta=0$, then X must be split, and it certainly follows that $X \oplus \langle \alpha\rangle X$ is split. But otherwise $\xi/\eta=-\alpha$ is a non-zero norm from X; therefore $X \cong \langle -\alpha\rangle X$ and it follows again that $X \oplus \langle \alpha\rangle X$ is split. This completes the proof. ∎

A special case of (4.3) which is particularly interesting is the case $\alpha_1=\cdots=\alpha_n=1$. The tensor product

$$(\langle 1\rangle \oplus \langle 1\rangle) \otimes \cdots \otimes (\langle 1\rangle \oplus \langle 1\rangle)$$

is then a 2^n-fold orthogonal sum of copies of $\langle 1\rangle$. We write this briefly as $2^n\langle 1\rangle$.

(4.4) Corollary. *For any field F and any $n \geqq 0$ the subset consisting of field element $\xi \neq 0$ which can be expressed as the sum of 2^n squares forms a multiplicative group.*

This follows from (4.2) and (4.3) since a field element is a norm from $2^n\langle 1\rangle$ if and only if it is a sum of 2^n squares. ∎

As an example, for the field **Q** of rational numbers, since both $5=1^2+2^2$ and $13=2^2+3^2$ can be expressed as the sum of 2 squares, it follows that $65(=4^2+7^2)$ can also. (Compare Chapter II, § 8.) By way of contrast, both $3=1^2+1^2+1^2$ and $5=0^2+1^2+2^2$ can be expressed as the sum of 3 squares, yet their product cannot. For if 15 were equal to $\alpha^2+\beta^2+\gamma^2$, then clearing denominators and reducing modulo 8 one would obtain

$$-d^2 \equiv a^2+b^2+c^2 \mod 8$$

with at least one of the integers a, b, c, d odd, which is easily seen to be impossible.

Here is an important application of (4.3).

Definition. If -1 is a sum of squares in F, then the *level* (Stufe) of F is the smallest integer s such that -1 is a sum of s squares. If -1 is not a sum of squares, we set $s=\infty$.

(4.5) Theorem. *For any field F, the order of the element $\langle 1\rangle$ in the additive group of $W(F)$ is precisely equal to $2s$. The level s is always either infinity or a power of 2.*

Remark. In the classical case of a number field, the level s is always equal to ∞, 1, 2, or 4. Examples are provided by $\mathbf{Q}$, $\mathbf{Q}(\sqrt{-1})$, $\mathbf{Q}(\sqrt{-2})$, and $\mathbf{Q}(\sqrt{-7})$ respectively. Since the order of $\langle 1\rangle$ is comparitively easy to compute, this theorem provides an excellent method for computing s.

Proof of (4.5). If $s=\infty$ the assertion is clear, so we may assume that $s<\infty$. Note first that the s-fold orthogonal sum $\langle 1\rangle\oplus\cdots\oplus\langle 1\rangle = s\langle 1\rangle$ is anisotropic. For if the equation

$$\xi_1^2+\cdots+\xi_s^2=0$$

had a solution with say $\xi_1\neq 0$, then it would follow that

$$-1=(\xi_2/\xi_1)^2+\cdots+(\xi_s/\xi_1)^2,$$

contradicting the definition of s. A similar argument shows that the $(s+1)$-fold orthogonal sum $(s+1)\langle 1\rangle$ is *not* anisotropic.

Now define an integer $n\geqq 0$ by the inequality $2^n\leqq s<2^{n+1}$. It follows a fortiori that the orthogonal sum $2^n\langle 1\rangle$ is anisotropic, but that $2^{n+1}\langle 1\rangle$ is not. Let us apply Theorem (4.3). Since $2^{n+1}\langle 1\rangle$ can be expressed as a tensor product

$$(\langle 1\rangle\oplus\langle 1\rangle)\otimes\cdots\otimes(\langle 1\rangle\oplus\langle 1\rangle),$$

and is not anisotropic, it follows that $2^{n+1}\langle 1\rangle$ is split. The relations

$$2^n\langle 1\rangle\nsim 0,\qquad 2^{n+1}\langle 1\rangle\sim 0$$

now clearly imply that the order of the element $\langle 1\rangle$ in the Witt ring is precisely equal to 2^{n+1}.

We must prove that $s=2^n$. Adding 2^n copies of $\langle -1\rangle$ to both sides of the relation $2^{n+1}\langle 1\rangle\sim 0$, we conclude that

$$2^n\langle 1\rangle\sim 2^n\langle -1\rangle.$$

But these spaces are anisotropic, so it follows from (1.7) that

$$2^n\langle 1\rangle\cong 2^n\langle -1\rangle.$$

Therefore -1 is a sum of 2^n squares, hence $s\leqq 2^n$, and therefore $s=2^n$. This completes the proof. □

(4.6) Remark. We can now give a more constructive proof that every element of the fundamental ideal $I \subset W(F)$ is nilpotent, whenever $s = 2^n < \infty$. (Compare § 3.6.) In fact for any w in the Witt ring, setting

$$w = \langle \alpha_1 \rangle + \cdots + \langle \alpha_r \rangle$$

we have

$$w^2 \equiv \langle \alpha_1 \rangle^2 + \cdots + \langle \alpha_r \rangle^2 = r \langle 1 \rangle \quad \mod 2\, W(F).$$

If the rank r is even, it follows that

$$w^2 \equiv 0 \quad \mod 2\, W(F)$$

and therefore

$$w^{2(n+1)} \equiv 0 \quad \mod 2^{n+1}\, W(F).$$

Since multiplication by $2^{n+1} = 2s$ annihilates every element of the Witt ring, this proves that $w^{2(n+1)} = 0$. □

Here is a concluding problem for the reader. An F-linear bijection $f\colon X \to X$ is called a *similarity transformation* if there exists a field element $\lambda \neq 0$ so that

$$f(x) \cdot f(y) = \lambda\, x \cdot y$$

for all x and y.

Exercise. Assume that 1 is a norm from X. Prove that the group of similarity transformations of X operates transitively on $X - 0$ if and only if X is multiplicative and anisotropic.

§ 5. The Powers of the Fundamental Ideal

This section will discuss the chain of ideals $I \supset I^2 \supset I^3 \supset \cdots$ in the Witt ring $W(F)$, where I is the fundamental ideal consisting of all Witt classes with even rank. A basic theorem concerning these ideals has recently been proved by Arason and Pfister. We state it here without proof.

(5.1) Theorem. *If $w \in I^n$ and $w \neq 0$, then $\|w\| \geqq 2^n$.*

(Here $\|w\|$ denotes the rank of the anisotropic representative for the Witt class w.) It follows immediately that the intersection of the ideals I^n is zero.

Each quotient I^n/I^{n+1} is clearly a vector space over the field

$$W(F)/I \cong \mathbf{F}_2 .$$

The following observation is due to Pfister.

(5.2) Theorem. *The quotient I/I^2 is canonically isomorphic to $F^{\bullet}/F^{\bullet 2}$.*

The proof is based on the determinant operation of Chapter I, § 2. For any inner product space X, recall that $\det(X)$ is a well defined element of $F^\bullet/F^{\bullet 2}$. However we have to be careful since the determinant of a split inner product space is not necessarily trivial. To correct this, we make the following modification.

Definition. For any inner product space X of rank r, the *discriminant*

$$d(X) \in F^\bullet/F^{\bullet 2}$$

is defined to be the element $(-1)^{r(r-1)/2}\det(X)$.

Let us compute the discriminant of an orthogonal sum $X \oplus Y$. Setting $r = \mathrm{rk}(X)$ and $s = \mathrm{rk}(Y)$, and setting $f(r) = r(r-1)/2$, the identity

$$f(r+s) = f(r) + f(s) + rs$$

is easily verified. It follows that

$$d(X \oplus Y) = (-1)^{rs} d(X)\, d(Y).$$

In particular, if either X or Y has even rank, then $d(X \oplus Y) = d(X)\, d(Y)$.

(5.3) Lemma. *The discriminant $d(X)$ depends only on the Witt class of X.*

For if S is split, of rank $2n$, then by Chapter I, § 6.3, the inner product matrix of S with respect to a suitable basis has the form $\begin{pmatrix} 0 & I \\ I & * \end{pmatrix}$. Therefore

$$\det(S) = (-1)^n F^{\bullet 2},$$

and it follows that $d(S)$ is the identity element of $F^\bullet/F^{\bullet 2}$. The proof of (5.3) is now straightforward. ∎

Proof of (5.2). Evidently the correspondence $w \mapsto d(w)$ maps the additive group of I homomorphically to the group $F^\bullet/F^{\bullet 2}$. This homomorphism is surjective since

$$d(\langle \xi \rangle + \langle -1 \rangle) = \xi F^{\bullet 2};$$

and it annihilates the ideal I^2 since I^2 is additively generated by products of the form

$$(\langle \alpha \rangle + \langle 1 \rangle)(\langle \beta \rangle + \langle 1 \rangle) = \langle \alpha\beta \rangle + \langle \alpha \rangle + \langle \beta \rangle + \langle 1 \rangle,$$

each of which has trivial discriminant. Now let $w = \langle \alpha_1 \rangle + \cdots + \langle \alpha_{2r} \rangle$ be an arbitrary element of I. Using the congruences

$$\langle \alpha \rangle + \langle \beta \rangle \equiv \langle -\alpha\beta \rangle + \langle -1 \rangle \quad \mod I^2$$

and

$$\langle -1 \rangle + \langle -1 \rangle + \langle -1 \rangle \equiv \langle 1 \rangle \quad \mod I^2,$$

we see by induction on r that w is congruent to an expression of the form $\langle\xi\rangle+\langle-1\rangle$ modulo I^2. Now if

$$d(w)=d(\langle\xi\rangle+\langle-1\rangle)=\xi F^{\bullet 2}$$

is the identity element of $F^\bullet/F^{\bullet 2}$, then $\langle\xi\rangle+\langle-1\rangle=0$, and it follows that $w\in I^2$. Thus the sequence $0\to I^2\to I\to F^\bullet/F^{\bullet 2}\to 1$ is exact; which completes the proof. □

The next step is naturally to look at the quotient I^2/I^3.

(5.4) Definition. A *symbol* on F with values in the group $\mathbf{Z}^\bullet=\{\pm 1\}$ is a bimultiplicative function

$$\varphi\colon\ F^\bullet\times F^\bullet\to\mathbf{Z}^\bullet$$

which satisfies the identity $\varphi(\alpha, 1-\alpha)=1$ for all $\alpha\neq 0, 1$.

The word "bimultiplicative" means that $\varphi(\alpha,\beta)$ is multiplicative as a function of α for fixed β and multiplicative as a function of β for fixed α. In particular $\varphi(\alpha, 1)=1$.

Remarks. The prototype for such an object is the "Hilbert symbol". If F is a local field of characteristic $\neq 2$, Hilbert showed that there is one and only one non-trivial symbol on F with values in $\mathbf{Z}^\bullet$. More recently, symbols with values in an arbitrary commutative group have arisen in *R*. Steinberg's analysis of central extensions of classical groups. Compare [Milnor, Introduction to Algebraic *K*-Theory].

(5.5) Lemma. *Given some fixed symbol φ on F with values in $\mathbf{Z}^\bullet$, the image $\varphi(\alpha,\beta)$ depends only on the Witt class of the inner product space $\langle\alpha\rangle\oplus\langle\beta\rangle$.*

Proof. Since the target group $\mathbf{Z}^\bullet$ has exponent 2, the image $\varphi(\alpha,\beta)$ is not changed if we multiply α or β by a square. Using the identity

$$-\alpha=(1-\alpha)/(1-\alpha^{-1}),$$

we see easily that

$$\varphi(\alpha, -\alpha)=1,$$

so that $\varphi(\alpha,\beta)$ is trivial whenever the inner product space $\langle\alpha\rangle\oplus\langle\beta\rangle$ is split.

Suppose then that the inner product spaces $\langle\alpha\rangle\oplus\langle\beta\rangle\sim\langle\gamma\rangle\oplus\langle\delta\rangle$ are not split. Then these spaces are isomorphic, so the equation

$$\gamma=\alpha\xi^2+\beta\eta^2$$

has a solution. Note the relation

$$\alpha\beta\equiv\gamma\delta \pmod{F^{\bullet 2}}.$$

If $\eta=0$, then $\alpha\equiv\gamma$, $\beta\equiv\delta \pmod{F^{\bullet 2}}$ so certainly $\varphi(\alpha,\beta)=\varphi(\gamma,\delta)$. If $\xi=0$, a similar argument shows that

$$\varphi(\gamma,\delta)=\varphi(\beta,\alpha).$$

But the symmetry relation

$$\varphi(\beta,\alpha)=\varphi(\alpha,\beta)^{-1}=\varphi(\alpha,\beta)$$

can easily be established by expanding the identity $\varphi(\alpha\beta,-\alpha\beta)=1$. Finally, suppose that $\xi\neq 0$ and $\eta\neq 0$. Then

$$\alpha\xi^2/\gamma+\beta\eta^2/\gamma=1$$

hence

$$\varphi(\alpha\gamma,\beta\gamma)=\varphi(\alpha\xi^2/\gamma,\beta\eta^2/\gamma)=1.$$

Expanding, it follows that

$$\varphi(\alpha,\beta)=\varphi(\gamma,\alpha\beta\gamma)$$

which completes the proof, since $\alpha\beta\gamma\equiv\delta$ modulo $F^{\bullet 2}$. ∎

(5.6) Lemma. *Suppose that the two inner product spaces*

$$\langle\alpha_1\rangle\oplus\cdots\oplus\langle\alpha_n\rangle \quad \textit{and} \quad \langle\beta_1\rangle\oplus\cdots\oplus\langle\beta_n\rangle$$

have the same rank and Witt class. Then it is possible to pass from the sequence $\alpha_1,\ldots,\alpha_n$ to the sequence $\beta_1,\ldots,\beta_n$ by changing just two entries at a time, preserving the Witt class at every stage.

The proof of this classical lemma will be deferred until the end of § 5.

Now for any inner product space X which possesses an orthogonal basis,

$$X\cong\langle\alpha_1\rangle\oplus\cdots\oplus\langle\alpha_n\rangle,$$

we defined the *Hasse invariant* $H_\varphi(X)\in\mathbf{Z}^\bullet$ to be the product $\prod_{i<j}\varphi(\alpha_i,\alpha_j)$. If X does not possess an orthogonal basis, then it must be symplectic, with F of characteristic 2, and we set $H_\varphi(X)=1$.

(5.7) Theorem. *The Hasse invariant $H_\varphi(X)$ does not depend on the choice of orthogonal decomposition. In fact if two spaces X and X' have the same rank and Witt class, then $H_\varphi(X)=H_\varphi(X')$. The identity*

$$H_\varphi(X\oplus Y)=H_\varphi(X)\,H_\varphi(Y)\,\varphi(\det X,\det Y)$$

is satisfied for all X and Y.

The proof, making use of (5.5) and (5.6), is straightforward and will be left to the reader. ∎

Let S be a split inner product space of rank $n=2m$. Then S has the same rank and Witt class as the m-fold sum $m(\langle 1\rangle\oplus\langle -1\rangle)$. It follows easily that

$$H_\varphi(S)=\varphi(-1,1)^{m(m-1)/2}.$$

This invariant is not always equal to 1; but if $n=2m\equiv 0 \bmod 8$, then certainly $H_\varphi(S)=1$.

Definition. For any Witt class w in the fundamental ideal I, the *Hasse-Witt invariant* $h_\varphi(w)$ is defined as follows. Choose a representative inner product space X for the Witt class w so that

$$\operatorname{rank}(X)\equiv 0 \pmod 8,$$

and set $h_\varphi(w)$ equal to the Hasse invariant $H_\varphi(X)$.

This function is well defined, since if $X\sim X'$ with $\operatorname{rank}(X)\equiv\operatorname{rank}(X')\equiv 0 \pmod 8$, then $X\oplus S\cong X'\oplus S'$ with $\operatorname{rank}(S)\equiv\operatorname{rank}(S')\equiv 0 \pmod 8$, and therefore

$$H_\varphi(X)=H_\varphi(X\oplus S)=H_\varphi(X'\oplus S')=H_\varphi(X').$$

(5.8) Theorem. *For each symbol φ the restriction of the Hasse-Witt function h_φ to the ideal I^2 yields a well defined homomorphism*

$$h_\varphi\colon I^2\to \mathbf{Z}^\bullet.$$

An element w of I^2 is annihilated by every one of these homomorphisms h_φ if and only if $w\in I^3$.

(Compare [Milnor, 1970].) In fact, the proof will show that the group of all symbols φ on F with values in $\mathbf{Z}^\bullet$ is canonically isomorphic to the group $\operatorname{Hom}(I^2/I^3,\mathbf{Z}^\bullet)$.

Proof of (5.8). The identity

$$h_\varphi(w+w')=h_\varphi(w)\,h_\varphi(w')\,\varphi\big(d(w),d(w')\big)$$

is evidently satisfied for any w and w' in the fundamental ideal I. If $w\in I^2$, so that $d(w)=1$, it follows that

$$h_\varphi(w+w')=h_\varphi(w)\,h_\varphi(w').$$

Next let us compute the Hasse-Witt invariant of a product $w\langle\beta\rangle$. Setting $w=\langle\alpha_1\rangle+\cdots+\langle\alpha_m\rangle$ with $m\equiv 0 \pmod 8$, we have

$$\begin{aligned}h_\varphi(w\langle\beta\rangle)&=\prod_{i<j}\varphi(\alpha_i\beta,\alpha_j\beta)=h_\varphi(w)\,\varphi(\alpha_1\ldots\alpha_m,\beta)^{m-1}\,\varphi(\beta,\beta)^{m(m-1)/2}\\&=h_\varphi(w)\,\varphi\big(d(w),\beta\big).\end{aligned}$$

For any $w'\in I$, setting $w'=\langle\beta_1\rangle+\cdots+\langle\beta_n\rangle$ with $n\equiv 0 \pmod 4$, we have

$$\begin{aligned}h_\varphi(w\,w')&=h_\varphi(w)^n\,\varphi\big(d(w),\beta_1\ldots\beta_n\big)\,\varphi\big(d(w),d(w)\big)^{n(n-1)/2}\\&=\varphi\big(d(w),d(w')\big).\end{aligned}$$

In particular, if $w \in I$ and $w' \in I^2$ then $h_\varphi(w\,w')=1$. Therefore each h_φ annihilates I^3, and gives rise to a homomorphism

$$h_\varphi\colon\ I^2/I^3 \to \mathbf{Z}^\bullet.$$

Conversely, given any homomorphism $g\colon\ I^2/I^3 \to \mathbf{Z}^\bullet$, let φ denote the symbol

$$\varphi(\alpha, \beta) = g\big((\langle\alpha\rangle - \langle 1\rangle)(\langle\beta\rangle - \langle 1\rangle)\big).$$

Using the congruence

$$\langle\alpha\,\alpha'\rangle - \langle 1\rangle \equiv \langle\alpha\rangle - \langle 1\rangle + \langle\alpha'\rangle - \langle 1\rangle \qquad \operatorname{mod} I^2$$

we see that φ is bimultiplicative; and if $\alpha+\beta=1$ then the isomorphism

$$\langle\alpha\rangle \oplus \langle\beta\rangle \cong \langle 1\rangle \oplus \langle\alpha\beta\rangle$$

implies that $\varphi(\alpha, \beta)=1$. Since the associated homomorphism

$$h_\varphi\colon\ I^2/I^3 \to \mathbf{Z}^\bullet$$

coincides with g on every generator $(\langle\alpha\rangle - \langle 1\rangle)\ (\langle\beta\rangle - \langle 1\rangle)$ of I^2, it follows that $h_\varphi = g$. Now if an element $w \in I^2$ is annihilated by every h_φ, then it is annihilated by every homomorphism from I^2/I^3 to $\mathbf{Z}^\bullet$, and it follows that $w \in I^3$. □

(5.9) The classical examples. If F is a finite field, then an elementary argument due to Steinberg shows that every symbol on F is trivial, hence the ideal $I^2 \subset W(F)$ is zero. Compare Chapter IV, § 1.5.

If F is a finite extension of the field of p-adic numbers, then there is precisely one non-trivial symbol on F with values in $\mathbf{Z}^\bullet$, hence I^2/I^3 is cyclic of order 2. In fact, the rank, determinant, and Hasse invariant form a complete system of invariants for an inner product space over F. (See [O'Meara, p. 170].) It follows easily that the ideal I^2 is cyclic of order 2, and that $I^3=0$.

Now suppose that F is a finite extension of the field of rational numbers. In this case, a complete invariant for an inner product space over F is provided by the rank, determinant, and total signature, together with the Hasse invariants associated with all of the various local completions of F ([O'Meara, p. 189]). For an element w in the ideal $I^3 \subset W(F)$, it follows easily that the signature $\sigma(w)$ provides a complete invariant. *In fact the homomorphism σ maps I^3 bijectively to the ideal* $8\mathbf{Z}^\Omega$. In particular in the totally imaginary case $\Omega=\emptyset$ it follows that $I^3=0$.

Note that the "length" of the chain of ideals $I \supset I^2 \supset I^3 \supset \cdots$ in the case of a number field is either 2 or ∞ according as -1 is a sum of squares or not. Here is a computation of this "length" in a quite different case.

Let F be a field of characteristic 2. Then the homomorphism $\xi \mapsto \xi^2$ maps F isomorphically onto a subfield F^2. The degree of F over F^2 is a number of the form 2^ι, where ι can take the values 0, 1, 2, ..., ∞. Note that F is perfect if and only if $\iota = 0$. If E is a finite extension of degree n over F, then E^2 has degree n over F^2, hence E has degree 2^ι over E^2. In other words this measure ι of imperfection is invariant under finite extensions of F. It is also invariant under arbitrary separable algebraic extensions. However ι increases by 1 under a simple transcendental extension. (For another characterization of ι see [Zariski and Samuel, p. 130].)

(5.10) Theorem. *If the field F of characteristic 2 has degree 2^ι over F^2, then the ideal $I^n \subset W(F)$ is zero for $n > \iota$ and non-zero for $n \leqq \iota$.*

For example if F is finite then $\iota = 0$, so the fundamental ideal $I = I^1$ itself is zero.

Proof. The ideal I^n is additively generated by products of the form

$$X = (\langle 1 \rangle \oplus \langle \alpha_1 \rangle) \otimes \cdots \otimes (\langle 1 \rangle \oplus \langle \alpha_n \rangle).$$

By Pfister's theorem (4.3), every such product is either anisotropic or split. But if $n > \iota$, then the 2^n products $\alpha_{j_1} \alpha_{j_2} \dots \alpha_{j_k}$, where $\{j_1, \dots, j_k\}$ ranges over all subsets of $\{1, \dots, n\}$, cannot be linearly independent over F^2. Hence X, being the orthogonal sum of the $\langle \alpha_{j_1} \dots \alpha_{j_k} \rangle$, cannot be anisotropic. Therefore $X \sim 0$, and it follows that $I^n = 0$.

Conversely, if $n \leqq \iota$ then we can inductively construct elements $\alpha_1, \dots, \alpha_n$ of F so that no α_{j+1} belongs to the subfield $F^2(\alpha_1, \dots, \alpha_j)$. It follows that the 2^n products $\alpha_{j_1} \dots \alpha_{j_k}$ *are* linearly independent over F^2. Hence the inner product space X above is anisotropic; and therefore $I^n \neq 0$. This completes the proof. (Compare [Milnor, Symmetric inner products in characteristic 2].) ☐

To conclude § 5, we prove the Lemma (5.6). Suppose that

$$\langle \alpha_1 \rangle \oplus \cdots \oplus \langle \alpha_n \rangle \sim \langle \beta_1 \rangle \oplus \cdots \oplus \langle \beta_n \rangle.$$

We must prove by induction on n that we can change the α_i two at a time, preserving the Witt class, so as to transform one sequence to the other. Certainly any permutation of the α_i can be obtained as a composition of permutation involving only two elements. To start the induction, the statement is certainly true for $n = 2$.

First suppose that the two spaces are anisotropic, and hence isomorphic. Then the equation

$$\beta_1 = \alpha_1 \xi_1^2 + \cdots + \alpha_n \xi_n^2$$

has a solution. Let k be the number of indices i for which $\xi_i \neq 0$. The proof will be based on a subsidiary induction on k.

If $k=1$, say $\beta_1=\alpha_1\xi_1^2$, then $\langle\alpha_1\rangle\cong\langle\beta_1\rangle$, hence $\langle\alpha_2\rangle\oplus\cdots\oplus\langle\alpha_n\rangle\sim\langle\beta_2\rangle\oplus\cdots\oplus\langle\beta_n\rangle$, and the conclusion follows by induction on n. If $k\geqq 2$, then say $\xi_1\neq 0$ and $\xi_2\neq 0$, and we may assume that the field element

$$\gamma=\alpha_1\xi_1^2+\alpha_2\xi_2^2$$

is non-zero. Hence $\langle\alpha_1\rangle\oplus\langle\alpha_2\rangle\cong\langle\gamma\rangle\oplus\langle\delta\rangle$ for some δ. Substituting γ and δ for α_1 and α_2, the conclusion follows by induction on k.

Suppose then that the two spaces are not anisotropic. Then the equation

$$0=\alpha_1\xi_1^2+\cdots+\alpha_n\xi_n^2$$

has a non-trivial solution. Let $k\geqq 2$ be the number of $\xi_i\neq 0$. We will prove by induction on k that the given sequence can be transformed into one with $\langle\alpha_1\rangle\oplus\langle\alpha_2\rangle\sim 0$, changing only two α_i at a time. If $k=2$, this is clear. If $k>2$, then say

$$\alpha_1\xi_1^2+\cdots+\alpha_k\xi_k^2=0$$

with $\alpha_1\xi_1^2+\alpha_2\xi_2^2=\gamma\neq 0$, hence $\langle\alpha_1\rangle\oplus\langle\alpha_2\rangle\cong\langle\gamma\rangle\oplus\langle\delta\rangle$. As before we can substitute γ and δ for α_1 and α_2. Therefore, by induction on k, we can transform the sequence $\alpha_1,\ldots,\alpha_n$ into one with $\langle\alpha_1\rangle\oplus\langle\alpha_2\rangle\sim 0$. Similarly we can transform $\beta_1,\ldots,\beta_n$ to a sequence with $\langle\beta_1\rangle\oplus\langle\beta_2\rangle\sim 0$. Then

$$\langle\alpha_3\rangle\oplus\cdots\oplus\langle\alpha_n\rangle\sim\langle\beta_3\rangle\oplus\cdots\oplus\langle\beta_n\rangle,$$

and the conclusion follows by induction on n. This completes the proof. $\square$

Chapter IV. Discrete Valuations and Dedekind Domains

The first section of this chapter defines the two residue class form homomorphisms associated with a discrete valuation of a field. Section 2 uses the second residue class form homomorphisms to compute the Witt ring $W(\mathbf{Q})$ of the rational numbers, and to give a new proof that $W(\mathbf{Z}) \cong \mathbf{Z}$. Section 3 applies a similar construction to an arbitrary Dedekind domain D with quotient field F, constructing an exact sequence

$$0 \to W(D) \to W(F) \to \oplus W(D/\mathfrak{p}),$$

where the direct sum extends over all maximal ideals $\mathfrak{p}$ of D. The final section applies this sequence to the special case of the integers in a number field.

§ 1. The Homomorphism ∂_v: $W(F) \to W(\bar{F})$

Recall that a *discrete valuation* v on a field F is a homomorphism from $F^\bullet$ onto the additive group $\mathbf{Z}$ satisfying

$$v(\alpha+\beta) \geqq \operatorname{Min}(v(\alpha), v(\beta))$$

for α, β, $\alpha+\beta \neq 0$. It is convenient to set $v(0) = +\infty$. The associated *valuation ring* $\mathfrak{O}$ consists of all $\alpha \in F$ with $v(\alpha) \geqq 0$. This ring has a unique maximal ideal $\mathfrak{P}$ consisting of all α with $v(\alpha) > 0$. The quotient $\bar{F} = \mathfrak{O}/\mathfrak{P}$ is called the *residue class field*. The image of any $u \in \mathfrak{O}^\bullet$ will be denoted by $\bar{u} \in \bar{F}^\bullet$.

We will construct an additive homomorphism ∂_v: $W(F) \to W(\bar{F})$, well defined up to multiplication by units of the form $\langle \bar{u} \rangle$ in $W(\bar{F})$. In order to define ∂_v, it is convenient to give a presentation of $W(F)$ by generators and relations.

(1.1) Lemma (Witt). *The additive group $W(F)$ is generated by the elements $\langle \alpha \rangle$, with $\alpha \in F^\bullet$, subject only to the following relations and their consequences*

(i) $\langle \alpha \rangle = \langle \alpha \xi^2 \rangle$ *for* $\xi \neq 0$,
(ii) $\langle \alpha \rangle + \langle -\alpha \rangle = 0$,
(iii) $\langle \alpha \rangle + \langle \beta \rangle = \langle \alpha + \beta \rangle + \langle \alpha\beta(\alpha+\beta) \rangle$
for $\alpha + \beta \neq 0$.

Proof. These three relations are clearly satisfied in $W(F)$, and the fact that every relation follows from these is an easy consequence of Chapter III, § 5.6. ∎

Now choose a *prime element* $\pi \in \mathfrak{O}$, that is an element π with $v(\pi) = 1$, so that $\pi\mathfrak{O} = \mathfrak{P}$. Then every element of $F^\bullet$ can be written uniquely as a product $\pi^i u$ with $u \in \mathfrak{O}^\bullet$.

(1.2) Lemma (Springer, Knebusch). *Fixing* π, *and fixing an integer* $k = 0$ *or* 1, *there is one and only one additive homomorphism*

$$\psi^k\colon\ W(F) \to W(\bar{F})$$

which maps each generator $\langle \pi^i u \rangle$ *to either* $\langle \bar{u} \rangle$ *or* 0 *according as* $i \equiv k$ (mod 2) *or* $i \not\equiv k$ (mod 2).

Proof. By (1.1) we need only verify that each of the defining relations for $W(F)$ maps to a valid relation in $W(\bar{F})$. Only the third relation presents any problem. It will be convenient to let the symbol ε_i stand for either 1 or 0 according as $i \equiv k$ or $i \not\equiv k$ (mod 2). If

$$\pi^h u_1 + \pi^i u_2 = \pi^j u_3,$$

then we must prove that

$$\varepsilon_h \langle \bar{u}_1 \rangle + \varepsilon_i \langle \bar{u}_2 \rangle = \varepsilon_j \langle \bar{u}_3 \rangle + \varepsilon_{h+i+j} \langle \bar{u}_1 \bar{u}_2 \bar{u}_3 \rangle$$

in $W(\bar{F})$. After dividing by a suitable power of π (and interchanging the roles of ψ^0 and ψ^1 if necessary) we may assume that two of the three numbers h, i, j are 0, and that the third is $\geqq 0$.

Case 1. If $h = i = j = 0$, then $\bar{u}_1 + \bar{u}_2 = \bar{u}_3$, and the required equation certainly follows.

Case 2. If $h > i = j = 0$, then $\bar{u}_2 = \bar{u}_3$ hence $\langle \bar{u}_1 \rangle = \langle \bar{u}_1 \bar{u}_2 \bar{u}_3 \rangle$, and the equation follows. The case $i > 0$ is completely analogous.

Case 3. If $0 = h = i < j$, then $\bar{u}_1 + \bar{u}_2 = 0$ hence

$$\langle \bar{u}_1 \rangle + \langle \bar{u}_2 \rangle = 0, \qquad \langle \bar{u}_3 \rangle + \langle \bar{u}_1 \bar{u}_2 \bar{u}_3 \rangle = 0,$$

and again the required equation follows. This completes the proof. ∎

Definition. These homomorphisms ψ^0 and ψ^1 from $W(F)$ to $W(\bar{F})$ are called the two *residue class form homomorphisms* associated with the valuation v. We will be particularly interested in ψ^1, and will use the alternative notation

$$\partial_v\colon\ W(F)\to W(\bar F)$$

for ψ^1. Note that ψ^0 is well defined, but that $\psi^1=\partial_v$ depends on the particular choice of π.

Let $\mathfrak{O}\subset F$ be the valuation ring associated with v, and let $W(\mathfrak{O})\to W(F)$ be the natural ring homomorphism.

(1.3) Lemma. *The composition* $W(\mathfrak{O})\to W(F)\xrightarrow{\psi^1} W(\bar F)$ *is zero.*

In fact we will see in § 3 that the sequence $0\to W(\mathfrak{O})\to W(F)\to W(\bar F)\to 0$ is exact.

Proof of (1.3). Since $\mathfrak{O}$ is a local ring, any inner product space over $\mathfrak{O}$ can easily be expressed as an orthogonal sum of inner product spaces of rank 1, with inner product matrix (u), and inner product spaces of rank 2 with inner product matrix of the form

$$\begin{pmatrix}\alpha & 1\\ 1 & \beta\end{pmatrix}$$

with $\alpha\equiv 0 \bmod \mathfrak{P}$. In the first case, the corresponding element $\langle u\rangle$ in $W(F)$ certainly satisfies $\psi^1\langle u\rangle=0$. In the second case, if $\alpha\neq 0$, the corresponding element in $W(F)$ can be written as a sum $\langle\alpha\rangle+\langle\alpha(\alpha\beta-1)\rangle$, with $\alpha\beta-1\equiv -1 \bmod \mathfrak{P}$. Evidently ψ^1 annihilates any such sum. Finally, if $\alpha=0$, then the given summand is split and hence corresponds to the zero element in the Witt ring. Thus each orthogonal summand maps to zero in $W(\bar F)$, and the conclusion follows. □

(1.4) Lemma. *Each homomorphism* $\psi^k\colon\ W(F)\to W(\bar F)$ *carries the ideal* $I^n(F)$ *onto* $I^{n-1}(\bar F)$ *for* $n\geqq 1$.

Proof. Inspection shows that the sum $\psi^0+\psi^1$ is a ring homomorphism which carries the ideal $I(F)$ to $I(\bar F)$, and hence carries $I^n(F)$ to $I^n(\bar F)$. Thus for any $w_n\in I^n(F)$ we have

$$\psi^1(w_n)\equiv -\psi^0(w_n)\quad \big(\mathrm{mod}\ I^n(\bar F)\big).$$

Suppose inductively that $\psi^0(w_n)\equiv 0\ \big(\mathrm{mod}\ I^{n-1}(\bar F)\big)$. Then for any $w\in I(F)$ we have

$$\begin{aligned}\psi^0(w\,w_n)&=\psi^0(w)\,\psi^0(w_n)+\psi^1(w)\,\psi^1(w_n)\\ &\equiv\psi^0(w)\,\psi^0(w_n)-\psi^1(w)\,\psi^0(w_n)\\ &\equiv 0\ \ \big(\mathrm{mod}\ I^n(\bar F)\big),\end{aligned}$$

since $\psi^0(w)-\psi^1(w)$ clearly belongs to $I(\bar F)$. This completes the induction. Since every generator $(\langle\bar 1\rangle+\langle\bar u_1\rangle)\dots(\langle\bar 1\rangle+\langle\bar u_{n-1}\rangle)$ of $I^{n-1}(\bar F)$ is the image, under either ψ^0 or ψ^1, of the generator

$$(\langle 1\rangle+\langle\pi\rangle)(\langle 1\rangle+\langle u_1\rangle)\dots(\langle 1\rangle+\langle u_{n-1}\rangle)$$

of $I^n(F)$, this proves the lemma. □

We will be particularly interested in residue class fields $\bar{F}$ which are finite. Let $\mathbf{F}_q$ denote the field with q elements.

(1.5) Lemma. *For any finite field* $\mathbf{F}_q$ *the ideal* $I(\mathbf{F}_q)$ *is either zero or cyclic of order 2 according as q is even or odd. The additive group of* $W(\mathbf{F}_q)$ *is either cyclic of order 2, cyclic of order 4, or non-cyclic of order 4 according as q is even,* $q\equiv 3 \pmod 4$, *or* $q\equiv 1 \pmod 4$.

Proof. Given $\alpha, \beta\in\mathbf{F}_q^\bullet$, the equation

$$\alpha\xi^2+\beta\eta^2=1$$

has a solution by Chapter II, § 3.3. Therefore

$$\langle\alpha\rangle\oplus\langle\beta\rangle\cong\langle 1\rangle\oplus\langle\alpha\beta\rangle,$$

and it follows that the ideal $I^2(\mathbf{F}_q)$ is zero. Using Chapter III, § 5.2, it follows that

$$I(\mathbf{F}_q)\cong\mathbf{F}_q^\bullet/\mathbf{F}_q^{\bullet 2},$$

where the group $\mathbf{F}_q^\bullet$ is cyclic of order $q-1$. Hence this ideal $I(\mathbf{F}_q)$ has order 1 or 2 according as $q-1$ is odd or even.

If $q\equiv 3 \pmod 4$, then -1 is not a square in $\mathbf{F}_q$ (compare Chapter II, § 8.1), hence $\langle -1\rangle\not\cong\langle 1\rangle$, so $\langle 1\rangle\oplus\langle 1\rangle\not\sim 0$ and it follows easily that $W(\mathbf{F}_q)\cong\mathbf{Z}/4\mathbf{Z}$. On the other hand if $q\not\equiv 3 \pmod 4$, then $\langle -1\rangle\cong\langle 1\rangle$, and it follows that $W(\mathbf{F}_q)$ is an algebra over $\mathbf{Z}/2\mathbf{Z}$. This completes the proof. □

§ 2. Computation of $W(\mathbf{Q})$

This section will completely describe the additive structure of the Witt ring $W(\mathbf{Q})$. Also it will give another proof that $W(\mathbf{Z})\cong\mathbf{Z}$, not making use of the Hasse-Minkowski theorem.

For each prime number p the p-adic valuation of $\mathbf{Q}$ gives rise to an additive homomorphism

$$\psi^1=\partial_p\colon\ W(\mathbf{Q})\to W(\mathbf{F}_p).$$

For any fixed w in $W(\mathbf{Q})$, it is clear that $\partial_p(w)=0$ for almost all p. Therefore we can aggregate these homomorphisms ∂_p into one homomorphism $\partial\colon\ W(\mathbf{Q})\to\bigoplus W(\mathbf{F}_p)$. Let i denote the unique ring homomorphism from $\mathbf{Z}$ to $W(\mathbf{Q})$.

(2.1) Theorem. *The sequence*

$$0 \to \mathbf{Z} \xrightarrow{i} W(\mathbf{Q}) \xrightarrow{\partial} \bigoplus W(\mathbf{F}_p) \to 0$$

is split exact.

Here the direct sum extends over all prime numbers p. (Thus in order to describe the most classical of Witt rings $W(\mathbf{Q})$ we are led to consider inner product spaces over finite fields, including the field $\mathbf{F}_2$ which is excluded in the classical theory.) For the structure of $W(\mathbf{F}_p)$, see (1.5).

Proof of (2.1). For each integer $k \geqq 1$, let L_k denote the subring of $W(\mathbf{Q})$ generated by the elements $\langle 1\rangle, \langle 2\rangle, \ldots, \langle k\rangle$. Then clearly

$$L_1 \subset L_2 \subset L_3 \subset \cdots$$

with union equal to $W(\mathbf{Q})$. The ring L_1 is evidently isomorphic to $\mathbf{Z}$. Note that $L_k = L_{k-1}$ unless k is prime.

Henceforth we will ignore the ring structure and think of L_k as an additive group.

(2.2) Lemma. *For each prime number p, the additive homomorphism ∂_p: $W(\mathbf{Q}) \to W(\mathbf{F}_p)$ induces an isomorphism*

$$L_p / L_{p-1} \to W(\mathbf{F}_p).$$

Proof. The homomorphism ∂_p clearly annihilates L_{p-1} and maps L_p onto $W(\mathbf{F}_p)$. We will need the following subsidiary lemma.

(2.3) Lemma. *If the numbers n_i and n satisfy $0<|n_i|<p$, $0<|n|<p$, and*

$$n_1 \ldots n_r \equiv n \pmod{p},$$

then

$$\langle p\, n_1 \ldots n_r\rangle \equiv \langle p\, n\rangle \pmod{L_{p-1}}.$$

Proof. First consider the case $r=2$. Then

$$n_1\, n_2 = n + k\, p$$

where

$$|k| \leqq ((p-1)^2 + (p-1))/p < p.$$

If $k=0$, there is nothing to prove. Otherwise, tensoring the isomorphism

$$\langle n\rangle \oplus \langle k\, p\rangle \cong \langle n_1\, n_2\rangle \oplus \langle n_1\, n_2\, n\, k\, p\rangle$$

with $\langle p\rangle$, we obtain

$$\langle p\, n\rangle \oplus \langle k\rangle \cong \langle p\, n_1\, n_2\rangle \oplus \langle n_1\, n_2\, n\, k\rangle$$

and hence $\langle p\, n\rangle \equiv \langle p\, n_1\, n_2\rangle \pmod{L_{p-1}}$. For any $n_3, \ldots, n_r$ which are less than p in absolute value, it follows that

$$\langle p\, n\, n_3 \ldots n_r\rangle \equiv \langle p\, n_1\, n_2\, n_3 \ldots n_r\rangle \pmod{L_{p-1}}.$$

A straightforward induction now completes the proof of (2.3). $\square$

Proof of (2.2). For each generator $\langle \bar{n} \rangle$ of $W(\mathbf{F}_p)$, we can choose a representative n with $|n| < p$, and lift $\langle \bar{n} \rangle$ to the generator $\langle p\,n \rangle$ of L_p/L_{p-1}. We will check that these lifted elements satisfy all of the defining relations for $W(\mathbf{F}_p)$. (Compare (1.1).) Thus if

$$n' \equiv n\,m^2 \qquad (\text{mod } p)$$

where n', n, m are less than p in absolute value, then it follows from (2.3) that

$$\langle p\,n' \rangle \equiv \langle p\,n\,m^2 \rangle = \langle p\,n \rangle \qquad (\text{mod } L_{p-1}).$$

Clearly $\langle p\,n \rangle + \langle p(-n) \rangle = 0$; and if $n_1 + n_2 = n$, where we may assume that n_1 and n_2 satisfy $-p < n_1 < 0 < n_2 < p$, then

$$\langle p\,n_1 \rangle \oplus \langle p\,n_2 \rangle \cong \langle p\,n \rangle \oplus \langle p\,n\,n_1\,n_2 \rangle.$$

Thus these elements $\langle p\,n \rangle$ modulo L_{p-1} satisfy all of the defining relations for $W(\mathbf{F}_p)$, and therefore generate a subgroup of L_p/L_{p-1} which maps isomorphically to $W(\mathbf{F}_p)$. Since it follows from (2.3) that these elements $\langle p\,n \rangle$ generate L_p modulo L_{p-1}, this completes the proof. □

Proof of Theorem (2.1). We will show by induction on k that the homomorphism

$$L_k \to \bigoplus_{p \leq k} W(\mathbf{F}_p)$$

is surjective, with kernel equal to $L_1 \cong \mathbf{Z}$. Certainly this statement is true for $k=1$. Suppose then that it is true for $k-1$. We may assume that k is prime. Given an element in the direct sum, by (2.2) there exists an element w in L_k whose image has the correct k-th coordinate. Subtracting the image of w, surjectivity follows by induction.

Similarly, if $w \in L_k$ maps to zero, then (2.2) implies that $w \in L_{k-1}$ and it follows by induction that $w \in L_1 \cong \mathbf{Z}$.

Now passing to the direct limit as $k \to \infty$, we see that the required sequence

$$0 \to \mathbf{Z} \to W(\mathbf{Q}) \to \bigoplus W(\mathbf{F}_p) \to 0$$

is exact. Using the signature homomorphism $W(\mathbf{Q}) \to \mathbf{Z}$, it follows that it is split exact. In fact the splitting is unique, since all of the groups $W(\mathbf{F}_p)$ are torsion. This completes the proof of (2.1). □

As one corollary of (2.1), we obtain a weak form of the Hasse-Minkowski theorem.

(2.4) Corollary. *If an element w in the Witt ring $W(\mathbf{Q})$ maps to zero in the Witt ring $W(\mathbf{Q}_p)$ of the p-adic numbers for every prime p, and also maps to zero in $W(\mathbf{R})$, then $w=0$.*

Proof. This follows since the homomorphism $\partial_p\colon W(\mathbf{Q}) \to W(\mathbf{F}_p)$ can be factored through $W(\mathbf{Q}_p)$. $\square$

(2.5) Corollary. *If I denotes the fundamental ideal in $W(\mathbf{Q})$, then I^3 is the free additive group generated by $8\langle 1\rangle$.*

Proof. If $w \in I^3(\mathbf{Q})$, then $\partial_p(w) \in I^2(\mathbf{F}_p) = 0$, by (1.4) and (1.5). Hence w is a multiple of $\langle 1\rangle$. In fact w must be a multiple of $8\langle 1\rangle$ since the signature homomorphism carries $I^n(\mathbf{Q})$ to $2^n\mathbf{Z}$. $\square$

Now let us give an alternative proof of Lemma (4.1) of Chapter II.

(2.6) Corollary. *Let X be any inner product space over $\mathbf{Z}$. Then the induced inner product space $\mathbf{Q} \otimes X$ over $\mathbf{Q}$ is isomorphic to an orthogonal sum of copies of $\langle 1\rangle$ and $\langle -1\rangle$.*

Proof. Let $\mathbf{Z}_{(p)} \subset \mathbf{Q}$ denote the valuation ring associated with the p-adic valuation of $\mathbf{Q}$. Since the natural homomorphism $W(\mathbf{Z}) \to W(\mathbf{Q})$ factors through $W(\mathbf{Z}_{(p)})$, it follows from (1.3) that the composition

$$W(\mathbf{Z}) \to W(\mathbf{Q}) \to \bigoplus W(\mathbf{F}_p)$$

is zero. Hence the image of $W(\mathbf{Z})$ in $W(\mathbf{Q})$ consists precisely of all positive or negative multiples of $\langle 1\rangle$. Together with Chapter I, § 7.4, this completes the proof. $\square$

In particular, if X is indefinite, it follows that there exists a non-zero vector y in $\mathbf{Q} \otimes X$ with $y \cdot y = 0$. Multiplying by a suitable positive integer m, this yields a non-zero vector $x = m\,y$ in the lattice X with $x \cdot x = 0$. Thus we have reproved Lemma (4.1) of Chapter II.

(2.7) Corollary. *The Witt ring $W(\mathbf{Z})$ is isomorphic to $\mathbf{Z}$.*

Proof. If X represents an element in the kernel of the natural homomorphism $W(\mathbf{Z}) \to W(\mathbf{Q})$, then certainly X contains a vector $x \neq 0$ with $x \cdot x = 0$. Proceeding as in Chapter II, § 2.2, we can decompose X as an orthogonal sum $X_0 \oplus X_0^{\perp}$ where X_0 has inner product matrix $\begin{pmatrix} 0 & 1 \\ 1 & * \end{pmatrix}$. It follows inductively that X is split. (Compare § 3.3 below.) Hence $W(\mathbf{Z})$ maps isomorphically to the ring $\mathbf{Z} \subset W(\mathbf{Q})$. $\square$

Combining (2.1) and (2.7), we see that the sequence

$$0 \to W(\mathbf{Z}) \to W(\mathbf{Q}) \to \bigoplus W(\mathbf{Z}/p\,\mathbf{Z}) \to 0$$

is exact. In this form, the sequence admits a significant generalization, which is discussed in the next section.

§ 3. Dedekind Domains

Let D be a Dedekind domain: that is a commutative ring without zero divisors in which any non-zero ideal can be expressed uniquely as a product of maximal ideals. The quotient field of D will be denoted by $F \supset D$. Every maximal ideal $\mathfrak{p} \subset D$ gives rise to a $\mathfrak{p}$-adic valuation on F with residue class field $D/\mathfrak{p}$, and hence to an associated homomorphism

$$\partial_{\mathfrak{p}}\colon W(F) \to W(D/\mathfrak{p}).$$

We will refer to such maximal ideals briefly as *primes*. Let X be an inner product space over F. Given finitely many elements $x_1, \ldots, x_k$, including a basis for X over F, we can form the D-submodule

$$L = D x_1 + \cdots + D x_k \subset X$$

spanned by $x_1, \ldots, x_k$.

Definition. Any such finitely generated D-submodule, containing a basis for X over F, is called a *lattice* (or a *D-lattice*) in X.

Given a lattice $L \subset X$, the *dual lattice* $L^{\#} \subset X$ is defined to be the set of all $x \in X$ such that $x \cdot l \in D$ for all $l \in L$.

Note that $L^{\#}$ is in fact a D-module, canonically isomorphic to $\operatorname{Hom}_D(L, D)$. For every D-linear map $L \to D$ extends uniquely to an F-linear map $X \to F$ which must have the form $x \mapsto x \cdot x_0$ for some unique $x_0 \in L^{\#}$. Using the theorem that every finitely generated torsion free module over D is projective[4], we see that L and thus also $L^{\#}$ are finitely generated and projective. Clearly $L^{\#}$ contains a basis for X over F.

(3.1) Theorem. *An inner product space X over F contains a lattice L which is self-dual, $L = L^{\#}$, if and only if the Witt class of X belongs to the kernel of the homomorphism*

$$\partial_{\mathfrak{p}}\colon W(F) \to W(D/\mathfrak{p})$$

for every maximal ideal $\mathfrak{p}$ of D.

Evidently L is self-dual if and only if the given inner product on X, when restricted to L, makes L into an inner product space over D.

Proof of (3.1). First suppose that there is only one maximal ideal $\mathfrak{p}$ in D, so that D is the valuation ring associated with the $\mathfrak{p}$-adic valuation of F. Let X be an inner product space over F, and suppose that

$$X \cong \langle \pi u_1 \rangle \oplus \cdots \oplus \langle \pi u_m \rangle \oplus \langle u_{m+1} \rangle \oplus \cdots \oplus \langle u_n \rangle.$$

[4] See for example [Cartan-Eilenberg] or [Milnor, Introduction to Algebraic K-Theory].

If the Witt class of X belongs to the kernel of $\partial_{\mathfrak{p}}$, then the inner product space

$$\langle \bar{u}_1 \rangle \oplus \cdots \oplus \langle \bar{u}_m \rangle$$

over $D/\mathfrak{p}$ is split. Hence this inner product space has inner product matrix of the form $\begin{pmatrix} 0 & I \\ I & * \end{pmatrix}$, with respect to a suitable basis. Lifting to the local ring D, it follows easily that the inner product space $\langle u_1 \rangle \oplus \cdots \oplus \langle u_m \rangle$ over D has inner product matrix of the form $\begin{pmatrix} A & I \\ I & B \end{pmatrix}$ with respect to a suitable basis, where each entry of A belongs to the ideal $\mathfrak{p}$. Hence the same statement applies to the inner product space $\langle u_1 \rangle \oplus \cdots \oplus \langle u_m \rangle$ over F. Tensoring with $\langle \pi \rangle$, it follows that the inner product space $\langle \pi u_1 \rangle \oplus \cdots \oplus \langle \pi u_m \rangle$ has inner product matrix

$$\begin{pmatrix} \pi A & \pi I \\ \pi I & \pi B \end{pmatrix}.$$

Dividing each of the first $m/2$ basis vectors by π, this inner product matrix is replaced by

$$\begin{pmatrix} \pi^{-1} A & I \\ I & \pi B \end{pmatrix}.$$

Evidently this is a matrix with entries in D whose determinant is a unit of D. Hence this modified basis spans the required self-dual lattice in $\langle \pi u_1 \rangle \oplus \cdots \oplus \langle \pi u_m \rangle$. Forming the direct sum with the obvious self-dual lattice in $\langle u_{m+1} \rangle \oplus \cdots \oplus \langle u_n \rangle$, we have the required self-dual lattice in X.

There is one special case not covered by this argument. If F has characteristic 2 and X is symplectic, then we cannot find an orthogonal basis for X. But in that case X is an orthogonal sum of hyperbolic planes, and clearly possesses a self-dual lattice.

Now suppose that the Dedekind domain D possesses more than one maximal ideal $\mathfrak{p}$. For each $\mathfrak{p}$, let $D_{\mathfrak{p}} \subset F$ be the associated valuation ring. Choosing a basis $e_1, \ldots, e_n$ for X, note that each inner product $e_i \cdot e_j$ belongs to $D_{\mathfrak{p}}$ for all but a finite number of primes $\mathfrak{p}$. Similarly the determinant $\det(e_i \cdot e_j)$ is a unit of $D_{\mathfrak{p}}$ for all but a finite number of primes. Thus there exists a finite set S of primes so that the $D_{\mathfrak{p}}$-lattice

$$D_{\mathfrak{p}} e_1 + \cdots + D_{\mathfrak{p}} e_n$$

is self-dual for all $\mathfrak{p} \notin S$.

Now suppose that $\partial_{\mathfrak{p}}(X) \sim 0$ for all $\mathfrak{p}$. Then for each $\mathfrak{p} \in S$ we can choose a $D_{\mathfrak{p}}$-lattice which is self-dual by the argument above. We will need the following.

(3.2) Lemma. *Let X be a vector space over F with basis $e_1, \ldots, e_n$. Given a $D_\mathfrak{p}$-lattice $L_\mathfrak{p}$ in X for each prime $\mathfrak{p}$, subject to the restriction that*

$$L_\mathfrak{p} = D_\mathfrak{p} e_1 + \cdots + D_\mathfrak{p} e_n$$

for all but a finite number of $\mathfrak{p}$, there exists one and only one D-lattice

$$L = \bigcap L_\mathfrak{p}$$

such that the $D_\mathfrak{p}$-lattice spanned by L is equal to $L_\mathfrak{p}$ for every $\mathfrak{p}$.

This is proved, for example, in [O'Meara, § 81:14].

Combining (3.2) with the discussion above, we construct a D-lattice L with the property that the induced $D_\mathfrak{p}$-lattice $D_\mathfrak{p} L$ is self-dual for every maximal ideal $\mathfrak{p}$. Thus if $x, y \in L$ then $x \cdot y \in D_\mathfrak{p}$ for every $\mathfrak{p}$, hence $x \cdot y \in D$. This proves that $L \subset L^\#$. Conversely, if $x \in L^\#$, then $x \cdot D_\mathfrak{p} L \subset D_\mathfrak{p}$ for every prime $\mathfrak{p}$, hence

$$x \in \bigcap (D_\mathfrak{p} L)^\# = \bigcap D_\mathfrak{p} L = L.$$

Therefore the lattice L is self-dual.

Conversely, if X contains a self-dual lattice, then it follows easily from (1.3) that the image of X in $W(D/\mathfrak{p})$ is zero for every $\mathfrak{p}$. This completes the proof of (3.1). ∎

(3.3) Corollary. *For any Dedekind domain D, the sequence*

$$0 \to W(D) \to W(F) \to \bigoplus W(D/\mathfrak{p})$$

is exact, where the direct sum extends over all non-zero prime ideals.

Proof. If an inner product space L over D corresponds to a split inner product space over F, we must prove that L itself is split. We will think of L as a self-dual lattice in the inner product space $X = F \otimes_D L$. Let $N \subset X$ be a subspace of half the dimension with $N \cdot N = 0$, so that $N = N^\perp$. Then the intersection $N \cap L$ is clearly a self-orthogonal subspace of L. This intersection is a direct summand of L since the quotient

$$L/(N \cap L) \subset X/N$$

is finitely generated and torsion-free, hence projective. It is equal to $(N \cap L)^\perp$, since if an element $x \in L$ is orthogonal to $N \cap L$ then it is orthogonal to all of N, and hence belongs to $N^\perp \cap L = N \cap L$.

Thus L is split, and the sequence $0 \to W(D) \to W(F)$ is exact. The rest of the proof is straightforward, using (3.1). ∎

Remark. In contrast to the situation in § 2, we do not assert that the homomorphism $\partial\colon W(F) \to \bigoplus W(D/\mathfrak{p})$ is necessarily surjective.

(3.4) Example. If F is a finite extension of the rational numbers, then the cokernel of ∂ can be computed as follows. Let $\mathscr{C}$ denote the ideal class group of the Dedekind domain $D \subset F$. Then there is a unique homomorphism

$$W(D/\mathfrak{p}) \to \mathscr{C}/\mathscr{C}^2$$

which carries each generator $\langle \bar{u} \rangle$ of $W(D/\mathfrak{p})$ to the ideal class of $\mathfrak{p}$ modulo $\mathscr{C}^2$. It is now not difficult to verify that the sequence

$$W(F) \xrightarrow{\partial} \oplus W(D/\mathfrak{p}) \to \mathscr{C}/\mathscr{C}^2 \to 0$$

is exact. (One uses Lemma (4.4) below, together with the discussion preceding (4.4), to show that every element of $\oplus I(D/\mathfrak{p})$ lifts back to $I^2(F)$. It is then only necessary to verify that the cokernel of the homomorphism from $I(F)/I^2(F) \cong F^\bullet/F^{\bullet 2}$ to $\oplus W(D/\mathfrak{p})/I(D/\mathfrak{p}) \cong \oplus \mathbf{Z}/2$ induced by ∂ is isomorphic to $\mathscr{C}/\mathscr{C}^2$.)

(3.5) Example. Let D be the ring $\mathbf{R}[x, y]/(x^2+y^2-1)$ consisting of all polynomial functions on the circle. Then each point $(\cos\theta, \sin\theta)$ on the unit circle gives rise to an ideal $\mathfrak{p}_\theta$ consisting of polynomials $f(x, y)$ which vanish at $(\cos\theta, \sin\theta)$. Clearly the quotient $D/\mathfrak{p}_\theta$ is the field of real numbers, hence $W(D/\mathfrak{p}_\theta) \cong \mathbf{Z}$. The associated homomorphism

$$\partial_\theta \colon W(F) \to W(D/\mathfrak{p}_\theta) \cong \mathbf{Z}$$

is well defined up to sign. Clearly ∂_{θ_0} carries each generator $\langle f \rangle$ of $W(F)$ to either ± 1 or 0 according as the function $f(\cos\theta, \sin\theta)$ changes sign or not as the variable θ passes through θ_0. A choice of sign for ∂_θ is equivalent to a choice of local orientation for the circle. Choosing orientations coherently, we obtain the relation

$$\sum_\theta \partial_\theta \langle f \rangle = 0$$

for every $f \in F^\bullet$. It is now not difficult to verify that the cokernel of

$$\partial \colon W(F) \to \oplus W(D/\mathfrak{p})$$

is infinite cyclic.

§ 4. Number Fields

Let F be a finite extension of the rational numbers, and let D be the ring of all algebraic integers in F. (See for example [Lang, p. 20].) Since the structure of $W(F)$ is well understood (compare Chapter III, § 5.9), we can use the exact sequence (3.3) to describe the ring $W(D)$.

First some notation. Let d be the number of dyadic primes in D (that is the number of prime ideals $\mathfrak{p}$ such that $D/\mathfrak{p}$ has characteristic 2). Let r be the number of embeddings of F in the real numbers and c the number of pairs of conjugate embeddings of F as a dense subset of the complex numbers. Thus $r+2c$ is the degree of F over $\mathbf{Q}$.

Two non-zero ideals $\mathfrak{a}$ and $\mathfrak{b}$ in D are called *strictly equivalent* if $\mathfrak{a}=\tau\mathfrak{b}$ for some field element τ which is *totally positive* (i.e., positive with respect to every embedding of F in $\mathbf{R}$). Let $\hat{\mathscr{C}}$ be the group of all strict equivalence classes of non-zero ideals. This is a finite extension of the usual ideal class group, which we denote by $\mathscr{C}$.

(4.1) Theorem[5]. *The radical $\mathfrak{N}_D$, consisting of all nilpotent elements in the Witt ring $W(D)$, is a finite group with order equal to 2^{c+d-1} multiplied by the number of elements of order $\leqq 2$ in $\hat{\mathscr{C}}$.*

Clearly an element of $W(D)$ is nilpotent if and only if its image in $W(F)$ is nilpotent. Comparing Chapter III, § 3.6 and § 3.8, it follows that there is an exact sequence which takes the form

$$0 \to \mathfrak{N}_D \to W(D) \to \mathbf{F}_2 \to 0$$

if F is totally imaginary ($r=0$); and an exact sequence

$$0 \to \mathfrak{N}_D \to W(D) \xrightarrow{\sigma} \mathbf{Z}^{\Omega}$$

otherwise. We will see in (4.5) that the image $\sigma(W(D))$ is always a subgroup of finite index in $\mathbf{Z}^{\Omega}$.

(4.2) Corollary. *If F is totally imaginary, then $W(D)$ is a finite ring with order equal to 2^{c+d} multiplied by the number of elements of order 2 in the ideal class group.*

Here are some examples. For the imaginary quadratic field $\mathbf{Q}(\sqrt{-2})$ or $\mathbf{Q}(\sqrt{-3})$, we have $W(D)\cong\mathbf{Z}/4\mathbf{Z}$; while for $\mathbf{Q}(\sqrt{-7})$, the ring $W(D)$ is isomorphic to $\mathbf{Z}/8\mathbf{Z}$. For $\mathbf{Q}(\sqrt{-1})$ it is non-cyclic of order 4 with basis $\langle 1\rangle$ and $\langle i\rangle$. Finally, for $\mathbf{Q}(\sqrt{-5})$, $W(D)$ is the sum of the ring $\mathbf{Z}/4\mathbf{Z}$ and an element of order 2 corresponding to an inner product space which is not free over D. (Compare the table below.)

(4.3) Corollary. *The radical $\mathfrak{N}_D$ is zero if and only if D is totally real, has only one dyadic prime, has odd class number, and contains units with arbitrarily prescribed sign. If these conditions are satisfied, then $W(D)$ is a free additive group with basis $\langle 1\rangle, \langle u_1\rangle, \ldots, \langle u_{r-1}\rangle$, where u_j is negative at the j-th ordering of F and positive at the remaining $r-1$ orderings.*

For example these conditions are satisfied for the fields $\mathbf{Q}$, $\mathbf{Q}(\sqrt{2})$, $\mathbf{Q}(\sqrt{5})$, and $\mathbf{Q}(\sqrt{13})$. Compare the table below, as well as the tables in [Borevich-Shafarevich].

The proofs of (4.2) and (4.3) are easily supplied. □

---

[5] We are grateful to Knebusch for pointing out an error in an earlier formulation of this result.

Table describing the additive structure of $W(D)$ for the quadratic field $\mathbf{Q}(\sqrt{n})$. The cyclic group of order m is denoted briefly by m.

$n=$	$W(D)\cong$	$n=$	$W(D)\cong$
2	$\mathbf{Z}\oplus\mathbf{Z}$	-1	$2\oplus 2$
3	$\mathbf{Z}\oplus\mathbf{Z}\oplus 2$	-2	4
5	$\mathbf{Z}\oplus\mathbf{Z}$	-3	4
6	$\mathbf{Z}\oplus\mathbf{Z}\oplus 2$	-5	$4\oplus 2$
7	$\mathbf{Z}\oplus\mathbf{Z}\oplus 2$	-6	$4\oplus 2$
10	$\mathbf{Z}\oplus\mathbf{Z}\oplus 2$	-7	8
11	$\mathbf{Z}\oplus\mathbf{Z}\oplus 2$	-10	$4\oplus 2$
13	$\mathbf{Z}\oplus\mathbf{Z}$	-11	4
14	$\mathbf{Z}\oplus\mathbf{Z}\oplus 2$	-13	$4\oplus 2$
15	$\mathbf{Z}\oplus\mathbf{Z}\oplus 2\oplus 2$	-14	$4\oplus 2\oplus 2$
		-15	$8\oplus 2$

These facts are easily verified, using [Borevich-Shafarevich, pp. 422–425] together with the proof of (4.1).

Proof of (4.1). We will think of $W(D)$ as a subring of $W(F)$. Hence we can intersect the radical $\mathfrak{N}_D$ with the ideal $I^2(F)$. As a first step in the proof, we will show that this intersection $\mathfrak{N}_D \cap I^2(F)$ is an elementary 2-group of order 2^{d-1}.

Let $F_\mathfrak{p}$ denote the $\mathfrak{p}$-adic completion of F. The classical theory, as described for example in [O'Meara], shows that an inner product space over $F_\mathfrak{p}$ is uniquely determined by its rank, determinant, and Hasse invariant. If the rank is $\geqq 3$, then the determinant in $F_\mathfrak{p}^\bullet/F_\mathfrak{p}^{\bullet 2}$ and the Hasse invariant in $\mathbf{Z}^\bullet$ can be prescribed arbitrarily. It follows easily that the ideal $I^2(F_\mathfrak{p})$ is cyclic of order 2, and that the Hasse-Witt homomorphism

$$h_\mathfrak{p}\colon I^2(F_\mathfrak{p}) \to \mathbf{Z}^\bullet$$

is bijective. (Compare Chapter III, §§ 5.4–5.9.)

Now recall that the homomorphism $\partial_\mathfrak{p}\colon I^2(F_\mathfrak{p}) \to I(D/\mathfrak{p})$ of § 1 is surjective. If the prime $\mathfrak{p}$ is not dyadic then the target group $I(D/\mathfrak{p})$ has two elements by (1.5), so it follows that the homomorphism

$$\partial_\mathfrak{p}\colon I^2(F_\mathfrak{p}) \to I(D/\mathfrak{p})$$

is also bijective. *Identifying $I(D/\mathfrak{p})$ with $\mathbf{Z}^\bullet$, this proves that the homomorphism*

$$\partial_\mathfrak{p}\colon I^2(F) \to I(D/\mathfrak{p}) \cong \mathbf{Z}^\bullet$$

can be identified with the $\mathfrak{p}$-th *Hasse-Witt invariant, for every prime* $\mathfrak{p}$ *which is not dyadic.*

The classical theory yields the following description of $I^2(F)$.

(4.4) Lemma. *An element w in $I^2(F)$ is uniquely determined by its Hasse-Witt invariants $h_{\mathfrak{p}}(w) \in \mathbf{Z}^{\bullet}$ for the various primes $\mathfrak{p}$, and by its signature $\sigma_P(w) \in 4\mathbf{Z}$ for the various orderings of F. These are subject only to the restriction that $h_{\mathfrak{p}}(w) = 1$ for almost all $\mathfrak{p}$, and that $\prod_{\mathfrak{p}} h_{\mathfrak{p}}(w)$ must be equal to $\prod_{P} (-1)^{\sigma_P(w)/4}$.*

Proof. This follows easily from the classical description of inner product spaces over F (compare [O'Meara, § 72]), together with the discussion in Chapter III, § 5. ∎

Now let us ask which elements w of $I^2(F)$ belong to the subring $W(D)$. A necessary and sufficient condition is that $\partial_{\mathfrak{p}}(w)$ must be zero for all $\mathfrak{p}$. We have seen that $\partial_{\mathfrak{p}}(w)$ can be identified with the $\mathfrak{p}$-th Hasse-Witt invariant whenever $\mathfrak{p}$ is non-dyadic; but that $\partial_{\mathfrak{p}} I^2(F) = 0$ when $\mathfrak{p}$ is dyadic. Evidently this proves the following.

(4.5) Corollary. *An element w in the intersection $W(D) \cap I^2(F)$ is uniquely specified by its Hasse-Witt invariants at the d dyadic primes, together with its signatures $\sigma_P(w) \in 4\mathbf{Z}$ at the r orderings of F. These are subject only to the relation $\prod_{\mathfrak{p}} h_{\mathfrak{p}}(w) = \prod_{P} (-1)^{\sigma_P(w)/4}$.*

In particular the image $\sigma(W(D) \cap I^2(F))$ is precisely equal to $4\mathbf{Z}^{\Omega}$, since there always exists at least one dyadic prime. *Therefore $\sigma(W(D))$ is a subring of finite index in $\mathbf{Z}^{\Omega}$.*

It follows that $W(D) \cap I^2(F)$ is the direct sum of a free abelian group of rank r, and an elementary 2-group of order 2^{d-1}. Restricting to the kernel of σ, *it follows that $\mathfrak{N}_D \cap I^2(F)$ is an elementary 2-group of order 2^{d-1}.*

Next we must study the quotient $\mathfrak{N}_D / \mathfrak{N}_D \cap I^2(F)$. Clearly this embeds in the quotient

$$I(F)/I^2(F) \cong F^{\bullet}/F^{\bullet 2},$$

and hence is an elementary 2-group. The precise image in $F^{\bullet}/F^{\bullet 2}$ can be identified as follows. Let $F^{\bullet}_{+} \subset F^{\bullet}$ denote the subgroup of index 2^r consisting of totally positive elements; and let $F^{\bullet}_{\mathrm{ev}}$ denote the subgroup consisting of elements α such that the $\mathfrak{p}$-adic value of α is even for all $\mathfrak{p}$. In other words $F^{\bullet}_{\mathrm{ev}}$ consists of all α such that the fractional ideal $D\alpha$ is equal to $\mathfrak{a}^2$ for some fractional ideal $\mathfrak{a}$.

(4.6) Lemma. *The quotient $\mathfrak{N}_D / \mathfrak{N}_D \cap I^2(F)$ is canonically isomorphic to the quotient $(F^{\bullet}_{+} \cap F^{\bullet}_{\mathrm{ev}})/F^{\bullet 2}$.*

Proof. We must decide which elements α of $F^{\bullet}$ modulo $F^{\bullet 2}$ can appear as the discriminants of inner product spaces over D with signature zero. If α is totally positive, and

$$D\alpha = \mathfrak{a}^2$$

for some fractional ideal $\mathfrak{a}$, then we can make $\mathfrak{a}$ into an inner product space over D by introducing the inner product

$$x \cdot y = x\,y/\alpha$$

for $x, y \in \mathfrak{a}$. Evidently $\mathfrak{a}$ is positive definite at P for every ordering P of F. Hence the inner product space

$$\mathfrak{a} \oplus \langle -1 \rangle$$

over D represents an element of $W(D) \subset W(F)$ with signature zero and discriminant $\alpha F^{\bullet 2}$.

Conversely, given an inner product space X of even rank and zero signature over D, we may assume (adding a hyperbolic plane if necessary) that the rank n is divisible by 4. The exterior power $\wedge^n X$ has rank 1 and is positive definite at every ordering. It is then easy to verify that $\wedge^n X$ is isomorphic to an ideal $\mathfrak{a}$ with inner product

$$x \cdot y = x\,y/\alpha$$

where α is totally positive, spans $\mathfrak{a}^2$, and where $d(X) = \alpha F^{\bullet 2}$. □

Now consider the inclusions

$$\begin{array}{ccc} & F^{\bullet 2} & \\ \nearrow & & \searrow \\ F_+^{\bullet 2} & & F_+^\bullet \cap F_{\mathrm{ev}}^\bullet \\ \searrow & & \nearrow \\ & D_+^\bullet F_+^{\bullet 2} & \end{array}$$

Let us assume first, to fix our ideas, that $r \geqq 1$. Going around the top of this diagram, since $F_+^\bullet$ has index 2^r in $F^\bullet$, it follows easily that $F_+^{\bullet 2}$ has index 2^{r-1} in $F^{\bullet 2}$. We have seen that the index of $F^{\bullet 2}$ in $F_+^\bullet \cap F_{\mathrm{ev}}^\bullet$ is equal to the order of $\mathfrak{N}_D/\mathfrak{N}_D \cap I^2(F)$.

Going around the bottom of the diagram, the Dirichlet unit theorem implies that the quotient

$$D_+^\bullet F_+^{\bullet 2}/F_+^{\bullet 2} \cong D_+^\bullet/D_+^{\bullet 2}$$

has order 2^{r+c-1} (still assuming that $r \geqq 1$). We claim that the quotient

$$(F_+^\bullet \cap F_{\mathrm{ev}}^\bullet)/D_+^\bullet F_+^{\bullet 2}$$

is canonically isomorphic to the group of elements of order 2 in $\mathscr{C}$. For each α in $F_+^\bullet \cap F_{\mathrm{ev}}^\bullet$ determines a unique fractional ideal $\mathfrak{a}$,

$$\mathfrak{a}^2 = D\alpha$$

representing an element of order 2 in $\hat{\mathscr{C}}$; and this ideal $\mathfrak{a}$ represents the identity element of $\hat{\mathscr{C}}$ if and only if α is the product of a unit and the square of a totally positive field element.

Now comparing the top and bottom of the diagram, we obtain the equality

$$2^{r-1}\,\mathrm{order}\,(\mathfrak{N}_D/\mathfrak{N}_D \cap I^2(F)) = 2^{r+c-1}\,\mathrm{order}\,(\hat{\mathscr{C}}_2)$$

providing that $r \geqq 1$. If $r=0$, a similar argument can be carried out, but the numbers 2^{r-1} and 2^{r+c-1} must be replaced by 1 and 2^c respectively. Thus in either case,

$$\mathrm{order}\,(\mathfrak{N}_D/\mathfrak{N}_D \cap I^2(F)) = 2^c\,\mathrm{order}\,(\hat{\mathscr{C}}_2).$$

Multiplying this equation by $\mathrm{order}\,(\mathfrak{N}_D \cap I^2(F)) = 2^{d-1}$, we obtain the required formula (4.1). ▯

Here is one final consequence of (4.1). Let D be the integers of an arbitrary number field F.

(4.7) Corollary. *The additive group of $W(D)$ is cyclic if and only if F is either $\mathbf{Q}$, or $\mathbf{Q}(\sqrt{-2})$, or $\mathbf{Q}(\sqrt{-p})$ where p is a prime of the form $4k+3$.*

Proof. If $W(D)$ is infinite cyclic, then it follows easily from (4.3) that $F=\mathbf{Q}$. Thus we may assume that $W(D)$ is finite, and hence that F is totally imaginary. Recall from the proof of (4.1) that the quotient

$$\mathfrak{N}_D/\mathfrak{N}_D \cap I^2(F)$$

is an elementary 2-group of order equal to 2^c times the number of elements of order 2 in the ideal class group. If $W(D)$ is cyclic, then evidently this quotient must be cyclic, hence c must be 1 and the class number of D must be odd.

Thus F is an imaginary quadratic field. An explicit formula for the class number of an imaginary quadratic field $\mathbf{Q}(\sqrt{-n})$ is given for example in [Borevich-Shafarevich, Ch. V, § 4.1]. Inspecting this formula, we see that the class number is odd if and only if the square-free integer n is equal to 1, 2, or a prime of the form $4k+3$. The cases $n=1$, 2, are easily handled, so we may assume that $n \geqq 3$.

If n is a prime of the form $8k+3$, there is only one dyadic prime in D (namely $\mathfrak{p}=2D$), so it follows from (4.2) that $W(D)$ has order 4. But -1 is not a square in F, so the element $\langle 1\rangle \in W(D)$ has order at least 4. This proves that $W(D)$ is cyclic of order 4.

On the other hand if n is a prime of the form $8k+7$, then the prime 2 splits in D, hence $W(D)$ has order 8. Using the Hasse-Witt invariant associated with one of the two dyadic primes, we see that

$$\langle 1\rangle \oplus \langle 1\rangle \oplus \langle 1\rangle \oplus \langle 1\rangle \not\sim 0,$$

so that $\langle 1\rangle$ has order at least 8. This shows that $W(D)$ is cyclic of order 8; and completes the proof. ▯

Chapter V. Some Examples

In this concluding chapter we briefly describe some examples of bilinear forms which arise naturally in topology, in differential geometry, and in number theory. The three sections of this chapter are completely independent.

§ 1. Homology Theory of Manifolds

It will be convenient to use the old fashioned terminology, and say that a manifold is *closed* if it is compact without boundary.

Let $M = M^{2n}$ be a closed manifold of dimension $2n$, and let $\mathbf{F}_2$ be the field with two elements. If x and y are homology classes in $H_n(M;\mathbf{F}_2)$, the *intersection number*

$$x \cdot y = y \cdot x \in \mathbf{F}_2$$

is defined. The Poincaré duality theorem, see for example [Spanier], implies that $H_n(M;\mathbf{F}_2)$ is an inner product space over $\mathbf{F}_2$, using the intersection number as inner product.

Let us look at the special case of a surface; that is let $n=1$. If M is connected then every homology class $x \in H_1(M;\mathbf{F}_2)$ can be represented by a simple closed curve $\gamma \subset M$. *Note that the self intersection number $x \cdot x$ is zero if and only if a small neighborhood N of γ is orientable.* For if N *is* orientable, then a small homotopy will deform γ to a curve $\gamma' \subset N$ which is disjoint from γ. (Compare Fig. 5.) But if γ does not possess an

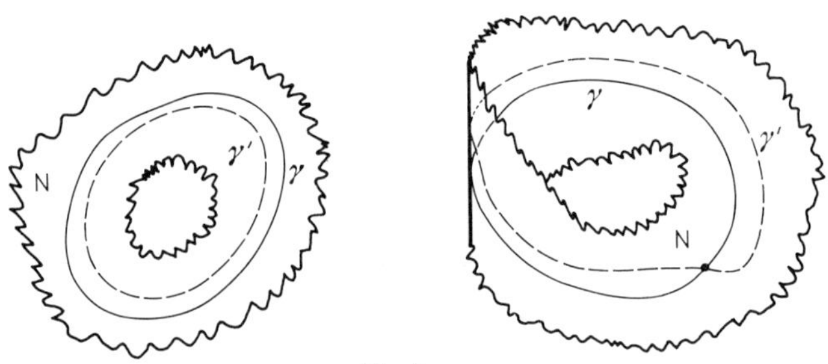

Fig. 5

orientable neighborhood, then it must possess a neighborhood N which is a Moebius band. In this case, deforming γ to a curve γ' which cuts γ transversally, the number of intersection points will be odd. This proves the following.

(1.1) Lemma. *The surface M is orientable if and only if $x \cdot x = 0$ for every $x \in H_1(M; \mathbf{F}_2)$.*

Since a closed connected surface is completely determined by its orientability or non-orientability, together with its middle Betti number, see for example [Massey], this implies:

(1.2) Corollary. *Two closed connected surfaces M and M' are homeomorphic if and only if the inner product spaces $H_1(M; \mathbf{F}_2)$ and $H_1(M', \mathbf{F}_2)$ are isomorphic.*

Here are some examples. For the torus T, the space $H_1(T; \mathbf{F}_2)$ is a hyperbolic plane, with basis e_1, e_2 and inner product matrix $\begin{pmatrix} 0 & 1 \\ 1 & 0 \end{pmatrix}$. (Compare Fig. 6.) For the projective plane P, there is just one basis vector e, with $e \cdot e = 1$, so that

$$H_1(P; \mathbf{F}_2) \cong \langle 1 \rangle .$$

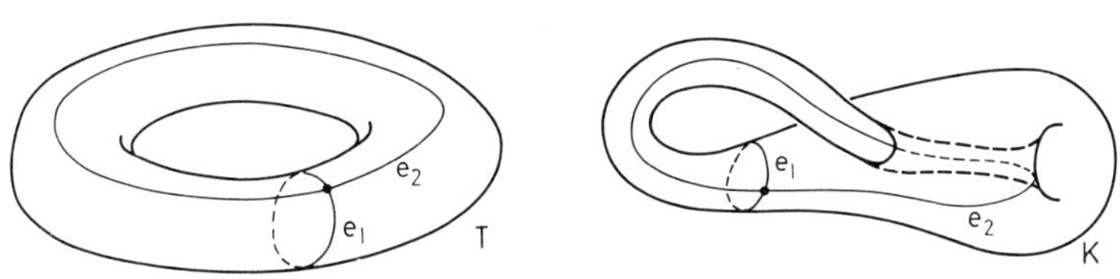

Fig. 6. The torus and Klein bottle

As a final example, for the Klein bottle K there is a basis e_1, e_2 with inner product matrix $\begin{pmatrix} 0 & 1 \\ 1 & 1 \end{pmatrix}$. Or alternatively, using the orthogonal basis $e_1 + e_2$, e_2, we see that

$$H_1(K; \mathbf{F}_2) \cong \langle 1 \rangle \oplus \langle 1 \rangle .$$

This decomposition corresponds of course to the geometric relation

$$K \cong P \# P,$$

where $\#$ denotes the "connected sum" operation: removing a small hole from each summand and then gluing boundaries together. Similarly the relation $\langle 1 \rangle \oplus \langle 1 \rangle \oplus \langle 1 \rangle \cong \langle 1 \rangle \oplus$(hyperbolic plane) of Chapter I, § 4, corresponds to the geometric relation.

$$P \# K \cong P \# T.$$

(1.3) Lemma. *Every inner product space over* $\mathbf{F}_2$ *is isomorphic to* $H_1(M; \mathbf{F}_2)$ *for some closed connected surface* M.

For by Chapter I, § 3.3 the given inner product space is isomorphic to

$$\langle 1 \rangle \oplus \cdots \oplus \langle 1 \rangle \oplus S$$

with S symplectic, and by I, § 3.5 every symplectic inner product space over $\mathbf{F}_2$ is isomorphic to an orthogonal sum of hyperbolic planes. Thus our inner product space can be realized as $H_1(M; \mathbf{F}_2)$ where M is a suitable connected sum $P \# \cdots \# P \# T \# \cdots \# T$. ▯

These results would remain true if we used any perfect field of characteristic 2 in place of $\mathbf{F}_2$.

The concept of a split inner product space is closely related to cobordism theory.

(1.4) Assertion. *The surface* M *bounds a compact* 3*-dimensional manifold if and only if the inner product space* $H_1(M; \mathbf{F}_2)$ *is split.*

In fact, more generally, if the closed manifold M^{2n} bounds a compact manifold P^{2n+1} then according to Thom the kernel N of the inclusion homomorphism

$$H_n(M^{2n}; \mathbf{F}_2) \to H_n(P^{2n+1}; \mathbf{F}_2)$$

satisfies $N \cdot N = 0$ and $\operatorname{rk}(N) = \frac{1}{2} \operatorname{rk} H_n(M^{2n}; \mathbf{F}_2)$. (Compare [Hirzebruch, § 8].) Therefore $N = N^{\perp}$, so that the inner product space $H_n(M^{2n}; \mathbf{F}_2)$ is split.

In the case of a surface the converse statement follows easily since the torus and Klein bottle are clearly boundaries. [Evidently, an inner product space over $\mathbf{F}_2$ is split if and only if its rank is even.] ▯

Remark. For some purposes it is better to work with a larger inner product space which contains all of the homology groups $H_i(M; \mathbf{F}_2)$. If $x \in H_i(M; \mathbf{F}_2)$ and $y \in H_j(M; \mathbf{F}_2)$, then the intersection number $x \cdot y$ is defined whenever $i + j = \dim M$. Setting $x \cdot y = 0$ if $i + j \neq \dim M$, the direct sum

$$H_{\oplus}(M; \mathbf{F}_2) = \bigoplus H_i(M; \mathbf{F}_2)$$

becomes an inner product space over $\mathbf{F}_2$. Clearly this inner product space is split if the dimension of M is odd, and is the orthogonal sum of $H_n(M; \mathbf{F}_2)$ and a split inner product space if the dimension is $2n$. The tensor product

$$H_{\oplus}(M; \mathbf{F}_2) \otimes H_{\oplus}(M'; \mathbf{F}_2)$$

is canonically isomorphic to the inner product space $H_{\oplus}(M \times M'; \mathbf{F}_2)$.

Now let M^{2n} be a closed *oriented* manifold, and let F be an arbitrary coefficient field. *Then $H_n(M^{2n}; F)$ is an inner product space over F, using the intersection number as inner product.* This inner product space is either symmetric or skew-symmetric according as n is even or odd. If M^{2n} bounds an oriented $(2n+1)$-manifold, then $H_n(M^{2n}; F)$ is a split inner product space.

Similarly one can work with integer coefficients. The **Z**-module $H_n(M^{2n}; \mathbf{Z})$/(torsion subgroup) is an inner product space over **Z**. The case of a simply-connected 4-manifold is particularly interesting.

(1.5) Theorem. *Let M and M' be closed oriented simply-connected 4-dimensional manifolds. There exists an orientation preserving homotopy equivalence $M \to M'$ if and only if the symmetric inner product space $H_2(M; \mathbf{Z})$ is isomorphic to $H_2(M'; \mathbf{Z})$.*

Remark. It is not known whether every inner product space over **Z** can be realized as $H_2(M; \mathbf{Z})$ for a suitable closed simply-connected 4-manifold M. In Chapter II we described a positive definite inner product space Γ_8 of rank 8 over **Z** satisfying

$$x \cdot x \equiv 0 \mod 2$$

for every $x \in \Gamma_8$. It would be extremely interesting to know whether $\Gamma_8 \cong H_2(M; \mathbf{Z})$ for some M. If such a manifold M exists, then by a theorem of Rohlin it cannot be given any piecewise-linear structure. (Compare [Milnor, 1958].)

Our special examples of surfaces have rather precise analogues among 4-manifolds. For example if P denotes the complex projective plane, then

$$H_2(P; \mathbf{Z}) \cong \langle 1 \rangle.$$

If $\bar{P}$ denotes the same manifold with reversed orientation, then

$$H_2(\bar{P}; \mathbf{Z}) \cong \langle -1 \rangle.$$

The analogue of the torus is the product $T = S^2 \times S^2$, having inner product matrix $\begin{pmatrix} 0 & 1 \\ 1 & 0 \end{pmatrix}$ with respect to a suitable basis e_1, e_2. Note that the diagonal homology class $e_1 + e_2$ has self intersection number $(e_1 + e_2) \cdot (e_1 + e_2)$ equal to 2. The analogue of a Klein bottle is the twisted S^2-bundle K over S^2. One can show by geometric arguments that

$$K \cong P \# \bar{P},$$

and that

$$K \# \bar{P} \cong T \# \bar{P},$$

even though the inner product space $H_2(K; \mathbf{Z})$ is clearly not isomorphic to $H_2(T; \mathbf{Z})$. (Compare Chapter I, § 4.)

Proof of Theorem (1.5). Let E denote the interior of an embedded 4-disk in M. Using the homology exact sequence of the pair $(M, M-E)$ with coefficients in $\mathbf{Z}$, together with Poincaré duality, we see easily that

$$H_i(M-E)=H_i(\text{point}) \quad \text{for } i\neq 2.$$

The group

$$H_2(M-E)\cong H_2(M)$$

is free abelian, say of rank r; and the Poincaré duality isomorphism

$$H_2(M)\cong H^2(M)\cong \operatorname{Hom}(H_2(M), \mathbf{Z})$$

shows that the intersection number is an inner product on $H_2(M)$.

Since $\pi_1(M-E)=0$, we have

$$\pi_2(M-E)\cong H_2(M-E)\cong H_2(M)$$

by the Hurewicz theorem. Hence there exists a map

$$f\colon S^2\vee\cdots\vee S^2\to M-E$$

from the r-fold bouquet of 2-spheres which induces a homology isomorphism

$$H_*(S^2\vee\cdots\vee S^2)\to H_*(M-E).$$

Since $M-E$ is an absolute neighborhood retract, it has the homotopy type of a CW-complex. Therefore, by a theorem of J.H.C. Whitehead, f is a homotopy equivalence. (Compare [Spanier, pp. 393–405].)

But M is obtained from $M-E$ by attaching a 4-cell. ([Spanier, p. 145].) Therefore M has the homotopy type of a space obtained from the r-fold bouquet $S^2\vee\cdots\vee S^2$ by attaching a 4-cell, using some attaching map

$$g\colon S^3\to S^2\vee\cdots\vee S^2.$$

We denote this resulting space by $(S^2\vee\cdots\vee S^2)\cup_g E^4$. The homotopy type of M is completely determined by the homotopy class of g in $\pi_3(S^2\vee\cdots\vee S^2)$.

To calculate the group $\pi_3(S^2\vee\cdots\vee S^2)$ we embed $S^2=P_1(\mathbf{C})$ in the infinite complex projective space $P_\infty(\mathbf{C})$, and use the inclusion

$$S^2\vee\cdots\vee S^2\subset S^2\times\cdots\times S^2\subset P_\infty(\mathbf{C})\times\cdots\times P_\infty(\mathbf{C}).$$

Denoting the r-fold bouquet by B, and the r-fold product of projective spaces by K, this inclusion $B\subset K$ gives rise to isomorphisms

$$H_4(K)\cong H_4(K, B)\cong \pi_4(K, B)\cong \pi_3(B).$$

In fact the first of these isomorphisms comes from the homology exact sequence of the pair (K, B), the second comes from the relative Hurewicz theorem, and the third from the homotopy exact sequence of the pair.

Clearly $H_4(K)$ is a free $\mathbf{Z}$-module. The dual module

$$H^4(K) \cong \operatorname{Hom}(H_4(K), \mathbf{Z})$$

has a basis consisting of products $u_i u_j$ with $i \leqq j$, where $u_1, \ldots, u_r$ form a basis for the cohomology group

$$H^2(K) \cong H^2(B) \cong H^2(M-E) \cong H^2(M).$$

([Spanier, p. 264].) Now extend the inclusion $B \to K$ to a mapping

$$B \underset{g}{\cup} E^4 \to K.$$

Lifting the products $u_i u_j \in H^4(K)$ back to $H^4(B \underset{g}{\cup} E^4) \cong H^4(M)$, and then evaluating on the orientation class $[M] \in H_4(M) \cong \mathbf{Z}$, we obtain a symmetric matrix of integers $u_i u_j [M]$ which completes describes the inner product space $H^2(M) \cong H_2(M)$. But evidently these integers also describe the homotopy class of the attaching map in $\pi_3(B) \cong \pi_4(K, B) \cong H_4(K)$. This completes the proof. ▯

§ 2. Rings of Smooth Real Valued Functions

The word *smooth* will mean k times continuously differentiable where k is fixed, $0 \leqq k \leqq \infty$.

Given a smooth paracompact manifold M, let $R(M)$ be the ring of all smooth real valued functions on M. Then a finitely generated projective module over $R(M)$ can be identified with the module $\Gamma(\xi)$ consisting of all smooth cross-sections of some real vector bundle ξ over M. (Compare [Swan, 1962].) The projective module $\Gamma(\xi)$ is free if and only if ξ is a trivial vector bundle.

An inner product on $\Gamma(\xi)$ (taking values in $R(M)$) is just a smooth function which assigns to each point x of M a real valued inner product on the fiber of ξ over x.

First consider the symplectic case. Vector bundles with symplectic inner product occur naturally in the theory of Hamiltonian differential equations. (See for example [MacLane].) Such bundles are classified by mappings of the base space M into a certain universal base space $BSp(2n, \mathbf{R})$ which has the same homotopy type as $BU(n)$. (Compare [Steenrod, § 19.8, 41.15]. Here $Sp(2n, \mathbf{R})$ denotes the group of isometries of a $2n$-dimensional real symplectic inner product space.) *Thus there is a one-to-one correspondence between symplectic inner product spaces of rank $2n$ over $R(M)$ and complex n-dimensional vector bundles over M.* ([Steenrod, § 41].) A symplectic inner product space possesses a symplectic basis if and only if the corresponding complex vector bundle is trivial.

One example which is particularly easy to understand is that of an oriented 2-dimensional vector bundle. Then $\wedge^2 \xi$ is a trivial line bundle, so the module $\Gamma(\xi)$ possesses a more or less unique symplectic inner product. If ξ is non-trivial (e.g., if ξ is the tangent bundle of the 2-sphere), then $\Gamma(\xi)$ certainly cannot possess a symplectic basis.

Now consider the symmetric case. Vector bundles with symmetric inner product occur naturally in Riemannian geometry, and in general relativity. The following statement is due to G. Lusztig. (Compare [Steenrod, § 40] as well as [Gel'fand, Mishchenko].)

Theorem. *The Witt ring $W(R(M))$, consisting of Witt classes of symmetric inner product spaces over $R(M)$, is canonically isomorphic to the ring $KO(M)$ of virtual real vector bundles over M.*

The proof can be sketched as follows. Let ξ be a real vector bundle over M with symmetric inner product. We will show first that ξ is isomorphic to an orthogonal sum $\xi^+ \oplus \xi^-$ where each fiber of ξ^+ is positive definite and each fiber of ξ^- is negative definite. For each fiber ξ_x of ξ let $P(\xi_x)$ be the set consisting of all positive definite subspaces of maximal rank in ξ_x. Then $P(\xi_x)$ has a natural topology, and in fact is a topological cell. To prove this, choose a base point $\eta_x^0 \in P(\xi_x)$. It is clear that the real inner product space ξ_x decomposes as

$$\xi_x = \eta_x^0 \oplus \eta_x^{0\perp}.$$

Now an arbitrary element $\eta_x \in P(\xi_x)$ can be described as the graph of a linear mapping $f \in \operatorname{Hom}(\eta_x^0, \eta_x^{0\perp})$ belonging to the convex open set described by the inequality

$$|f(v) \cdot f(v)| < v \cdot v$$

for all $v \in \eta_x^0$, $v \neq 0$.

Evidently these cells $P(\xi_x)$ form the fibers of a new fiber bundle over M. This new bundle possesses a cross-section

$$x \mapsto \xi_x^+ \in P(\xi_x)$$

(compare [Hirzebruch, p. 51]), and the resulting spaces ξ_x^+ form the fibers of the required sub-bundle $\xi^+ \subset \xi$ with $\xi = \xi^+ \oplus \xi^{+\perp} = \xi^+ \oplus \xi^-$ as required.

The two vector bundles ξ^+ and ξ^- are uniquely determined, up to isomorphism. For if ξ also decomposes as $\eta^+ \oplus \eta^-$ then $\eta^+ \cap \xi^- = 0$, hence the composition

$$\eta^+ \subset \xi \to \xi / \xi^- \cong \xi^+$$

is an isomorphism of vector bundles.

It is not difficult to show that the inner product space $\Gamma(\xi^+ \oplus \xi^-)$ splits if and only if the vector bundle ξ^+ is isomorphic to ξ^-. Further details will be left to the reader.

§ 3. The Discriminant of a Field Extension

Let $F \subset F'$ be finite extensions of the field of rational numbers. Then the ring R consisting of all algebraic integers in F is a Dedekind domain [Lang, p. 20], as is the ring R' consisting of all algebraic integers in F'. Forgetting its ring structure, we will think of the larger ring R' as a module over R. This R-module is torsion free and finitely generated [Lang, p. 6]. Therefore, by a classical theorem of Steinitz, R' is projective, being R-linearly isomorphic to a direct sum of the form $R \oplus \cdots \oplus R \oplus \mathfrak{a}$ where $\mathfrak{a}$ is a non-zero ideal in R. (See for example [Milnor, Introduction to Algebraic K-Theory, p. 11].) Evidently the number n of summands (including $\mathfrak{a}$) is equal to the degree of F' over F.

This module R' possesses a canonical R-valued symmetric bilinear form

$$\beta(x, y) = \operatorname{trace}_{F'/F}(x\,y).$$

Thus the pair (R', β) is a bilinear form space over R.

As in Chapter I, § 5.7, we will consider the n-th exterior power $\wedge_R^n R'$. This exterior power is a projective module of rank 1, being R-linearly isomorphic to the ideal $\mathfrak{a}$. It is provided with a canonical symmetric bilinear form $\hat{\beta}$, defined by

$$\hat{\beta}(x_1 \wedge \cdots \wedge x_n, y_1 \wedge \cdots \wedge y_n) = \det\big(\beta(x_i, y_j)\big).$$

The "discriminant" of the extension $R' \supset R$ is usually defined to be a certain ideal $\mathfrak{d}$ in R. ([Lang, p. 65].) For example $\mathfrak{d}$ can be described as the ideal generated by the image of the function

$$\hat{\beta}\colon \wedge_R^n R' \times \wedge_R^n R' \to R.$$

We obtain a slightly sharper invariant by defining the *discriminant* of R' over R to be the isomorphism class of the bilinear form space $(\wedge_R^n R', \hat{\beta})$ over R. (Compare [Fröhlich, 1960].)

As an example, suppose that R' happens to be free over R with basis $e_1, \ldots, e_n$. Then $\wedge_R^n R'$ is free over R with a single basis element $e = e_1 \wedge \cdots \wedge e_n$. Setting

$$d_0 = \hat{\beta}(e, e) = \det\big(\operatorname{trace}(e_i\, e_j)\big) \in R,$$

it follows that $(\wedge_R^n R', \hat{\beta})$ is isomorphic to the free bilinear form space $\langle d_0 \rangle = \langle d_0 \rangle_R$. This ring element d_0 is well defined up to multiplication by squares of units. Evidently the discriminant ideal $\mathfrak{d}$, generated by the image of $\hat{\beta}$, is equal to the principal ideal $d_0 R$.

Even if R' is not free over R, the field F' is certainly free over F. Working over F, the corresponding exterior power $\wedge_F^n F'$ is provided with a corresponding symmetric bilinear form $\hat{\beta}$. Evidently

$$(\wedge_F^n F', \hat{\beta}) \cong \langle d \rangle_F$$

for some field element $d \neq 0$, where d is well defined up to multiplication by $F^{\bullet 2}$. The following result is due to [Artin, 1950] and [Fröhlich, 1960].

(3.1) Lemma. *With d and $\mathfrak{d}$ as above, there is one and only one fractional ideal $\mathfrak{a}$ in F satisfying the equation*

$$\mathfrak{d} = d\mathfrak{a}^2. \tag{1}$$

The exterior power $\wedge_R^n R'$ is R-linearly isomorphic to this ideal $\mathfrak{a}$. Hence the ring R' is free over R if and only if $\mathfrak{a}$ is a principal ideal.

In fact the proof will show that the bilinear form space $(\wedge_R^n R', \hat{\beta})$ is isomorphic to $(\mathfrak{a}, \beta_d)$, where

$$\beta_d(a_1, a_2) = d\, a_1\, a_2 \tag{2}$$

for all $a_1, a_2 \in \mathfrak{a}$. We have already noted that R' is R-linearly isomorphic to $R \oplus \cdots \oplus R \oplus \mathfrak{a}_1$, and therefore $\wedge_R^n R' \cong \mathfrak{a}_1$, for some ideal $\mathfrak{a}_1$. Evidently any R-valued bilinear form on $\mathfrak{a}_1$ is equal to the form β_{d_1} defined by Eq. (2), for some $d_1 \in \mathfrak{a}_1^{-2}$. Thus

$$(\wedge_R^n R', \hat{\beta}) \cong (\mathfrak{a}_1, \beta_{d_1}), \tag{3}$$

and it follows that $\mathfrak{d} = d_1\, \mathfrak{a}_1^2$. Tensoring the isomorphism (3) with F, we see that $d_1 = d f^2$ for some $f \in F^\bullet$. Now defining $\mathfrak{a} = f\mathfrak{a}_1$, it follows that $\mathfrak{d} = d\mathfrak{a}^2$ and that

$$(\mathfrak{a}_1, \beta_{d_1}) \cong (\mathfrak{a}, \beta_d),$$

which completes the proof. □

(3.2) Example. *Suppose that $R = \mathbf{Z}[\sqrt{-5}]$ is the ring of integers in the field $\mathbf{Q}(\sqrt{-5})$, and that R' is the ring of integers in the quadratic extension field $\mathbf{Q}(\sqrt{-5}, \sqrt{2})$. Then the discriminant $(\wedge_R^2 R', \hat{\beta})$ of this extension is isomorphic to $(\mathfrak{p}, \beta_2)$ where $\mathfrak{p}$ denotes the prime ideal $2R + (1+\sqrt{-5})R$ with $\mathfrak{p}^2 = 2R$, and where $\beta_2(a_1, a_2) = 2a_1 a_2$ for $a_1, a_2 \in \mathfrak{p}$. The ideal $\mathfrak{p}$ is not principal, so the ring R' is not free over R.*

Proof. R' can be described more explicitly by noting that the elements $1, \sqrt{-5}, \sqrt{2}, \frac{1}{2}(\sqrt{2}+\sqrt{-10})$ form a basis for R' over $\mathbf{Z}$. This can be established, for example, by embedding $F' = \mathbf{Q}(\sqrt{-5}, \sqrt{2})$ in the cyclotomic field $\mathbf{Q}(e^{2\pi i/40})$, and using the fact that R' is the intersection of F'

with the ring $\mathbf{Z}[e^{2\pi i/40}]$ of cyclotomic integers. ([Lang, p. 75].) Details of the computation will be left to the reader. The ideal $\mathfrak{d}=\mathfrak{d}_{R'/R}$ can now be computed directly. Alternatively, one can first compute the two discriminants

$$(\wedge_{\mathbf{Z}}^{2} R, \hat{\beta}) \cong \langle -20\rangle_{\mathbf{Z}}, \qquad (\wedge_{\mathbf{Z}}^{4} R', \hat{\beta}) \cong \langle 6400\rangle_{\mathbf{Z}}$$

and then make use of the identity (where n is the degree of F' over F)

$$\mathfrak{d}_{R'/\mathbf{Z}} = \mathfrak{d}_{R/\mathbf{Z}}^{n} \operatorname{norm}_{R/\mathbf{Z}}(\mathfrak{d}_{R'/R}),$$

which follows from [Lang, pp. 60, 66]. In any case, one finds that the ideal $\mathfrak{d}_{R'/R}$ is equal to $\mathfrak{p}^4 = 4R$.

Choosing a basis e_1, e_2 for $F' = F(\sqrt{2})$ over F, a short computation shows that

$$\det(\operatorname{trace}(e_i e_j)) \equiv 2 \pmod{F^{\bullet 2}}.$$

Thus we can choose $d=2$ in Lemma (3.1). The equation $\mathfrak{d} = 2\mathfrak{a}^2$ now implies that $\mathfrak{a} = \mathfrak{p}$, which completes the proof. ∎

A different example, similar to this one, is described in [MacKenzie-Scheuneman].

Appendix 1. Quadratic Forms

The theory of symmetric bilinear forms is intimately related to the theory of quadratic forms. In fact, over a ring in which 2 is a unit, the two theories are indistinguishable. For this reason, it seems advisable to give a brief description of quadratic forms, even though our emphasis is on bilinear forms. Let X be an R-module. As usual we assume that the ring R is commutative with 1.

Definition. A *quadratic form* on X is a function $q: X \to R$ such that

$$q(\alpha x) = \alpha^2 q(x)$$

for all $\alpha \in R$, and such that the function $(x|y)$ on $X \times X$ defined by

$$(x|y) = q(x+y) - q(x) - q(y)$$

is bilinear over R.

As an example, if β is a bilinear form (not necessarily symmetric), then the function

$$q(x) = \beta(x, x)$$

is clearly quadratic, with associated symmetric bilinear form

$$(x|y) = \beta(x, y) + \beta(y, x).$$

If X is projective, then every quadratic form on X can be obtained in this way from a (non-symmetric) bilinear form. Two bilinear forms β and β' give rise to the same quadratic form if and only if the difference $\beta - \beta'$ is symplectic.

Note the identity $(x|x) = 2q(x)$. Thus the symmetric bilinear form $(x|y)$ associated with a quadratic form must satisfy the congruence

$$(x|x) \equiv 0 \mod 2R$$

for all x. If 2 is not a zero-divisor in R, then any symmetric bilinear form satisfying this congruence is associated with a unique quadratic form

$$q(x) = \tfrac{1}{2}(x|x).$$

In particular, if 2 is a unit then every symmetric bilinear form $(x|y)$ comes from a unique quadratic form $\frac{1}{2}(x|x)$.

Definition. Let $(X_1, q_1), \ldots, (X_n, q_n)$ be modules with quadratic forms over any commutative ring R. The *orthogonal sum* $X_1 \oplus \cdots \oplus X_n$ is defined to be the direct sum of the modules X_i with quadratic form q defined by the equation

$$q(x_1 \oplus \cdots \oplus x_n) = \sum q_i(x_i)$$

summed over $1 \leqq i \leqq n$.

The associated bilinear form $(x|y)$ on $X = X_1 \oplus \cdots \oplus X_n$ is the orthogonal sum of the associated bilinear forms $(x|y)_i$ on the X_i. Hence $(x|y)$ is an inner product if and only if each $(x|y)_i$ is an inner product.

Definition. We will call the pair (X, q) a *quadratic inner product space*, if X is finitely generated projective, and if the bilinear form $(x|y)$ associated with the quadratic form q is an inner product on X (Chapter I, § 1.1).

We observed in Chapter I, § 4 that the Witt Cancellation Theorem (4.4) is not true for inner product spaces in characteristic 2. It is interesting to note that the analogue of (4.4) for *quadratic* inner product spaces is true in any characteristic. That is, if

$$(X_1, q_1) \oplus (X, q) \cong (X_2, q_2) \oplus (X, q),$$

where (X_1, q_1), (X_2, q_2), and (X, q) are quadratic inner product spaces over a field, then $(X_1, q_1) \cong (X_2, q_2)$. This is proved in [Chevalley, p. 16] or [Bourbaki, p. 71].

The following basic remark is due to [Fröhlich, 1969] and independently to [Sah]. *If X_1 is a symmetric bilinear form module and X_2 is a quadratic form module, then the tensor product $X_1 \otimes X_2$ is a quadratic form module.* In fact, given a symmetric bilinear form β_1 on X_1 and a quadratic form q_2 on X_2, there is a unique quadratic form q on $X_1 \otimes X_2$ satisfying the equations, both being necessary for the definition,

$$q(x_1 \otimes x_2) = \beta_1(x_1, x_1)\, q_2(x_2)$$

and

$$(x_1 \otimes x_2 | y_1 \otimes y_2) = \beta_1(x_1, y_1)\, (x_2 | y_2).$$

Note the isomorphism

$$\langle 1 \rangle \otimes X_2 \cong X_2.$$

If both X_1 and X_2 are quadratic form modules, then using the associated bilinear form $(x_1|y_1)$ as β_1 it follows that $X_1 \otimes X_2$ is also a quadratic form module. The quadratic form q on $X_1 \otimes X_2$ is determined by the equations

$$q(x_1 \otimes x_2) = 2 q_1(x_1)\, q_2(x_2)$$

and

$$(x_1 \otimes x_2 | y_1 \otimes y_2) = (x_1|y_1)\, (x_2|y_2).$$

The factor of 2 is surprising but necessary. This factor was incorrectly left out in [Bourbaki, v. **24**, p. 137].

The *Witt algebra* $WQ(R)$ of quadratic inner product spaces over R can now be defined as follows. (Compare [Bass], [Sah].) A quadratic inner product space (X, q) is said to be *split* if the module X contains a direct summand N with $N = N^{\perp}$ and $q(N) = 0$. Two quadratic inner product spaces (X_1, q_1) and (X_2, q_2) belong to the same *Witt class* if

$$(X_1, q_1) \oplus (S_1, q_1') \cong (X_2, q_2) \oplus (S_2, q_2')$$

where the (S_i, q_i') are split. The Witt classes of quadratic inner product spaces over R now form the required algebra $WQ(R)$. Clearly $WQ(R)$ is a commutative associative algebra over the Witt ring $W(R)$. This algebra does not possess a 1 element in general. There is a canonical augmentation

$$a\colon\ WQ(R) \to W(R),$$

and the product of two arbitrary elements w_1 and w_2 in $WQ(R)$ is determined by the identity $w_1 w_2 = a(w_1) w_2$. Of course if 2 is a unit in R, then a is an isomorphism.

In the case of a field F, it is easily verified that every quadratic inner product space over F is the orthogonal sum of a split quadratic inner product space and an anisotropic quadratic inner product space (i.e., one with $Q(x) \neq 0$ for $x \neq 0$). Furthermore, any two split spaces of the same rank are isomorphic. Using the Witt cancellation theorem, it follows that *every Witt class in $WQ(F)$ possesses one and up to isomorphism only one anisotropic representative.*

Suppose in particular that F is a field of characteristic 2. Then the bilinear form $(x|y)$ associated with a quadratic form is necessarily symplectic

$$(x|x) = 2q(x) = 0.$$

Therefore the augmentation homomorphism

$$WQ(F) \to W(F)$$

is identically zero. For if (X, q) is a quadratic inner product space, then X possesses a symplectic basis (Chapter I, § 3.5), and hence represents the zero element of the Witt ring $W(F)$.

An important invariant of a quadratic inner product space in characteristic 2 has been defined by C. Arf. Let

$$\wp\colon\ F \to F$$

denote the additive homomorphism $\wp(\xi) = \xi^2 + \xi$. Choosing a symplectic basis $x_1, \ldots, x_n$, $y_1, \ldots, y_n$ for X, with inner product matrix $\begin{pmatrix} 0 & I \\ I & 0 \end{pmatrix}$, Arf showed that the residue class of

$$q(x_1)\, q(y_1) + \cdots + q(x_n)\, q(y_n)$$

modulo $\wp(F)$ is an invariant of (X, q). This Arf invariant depends only on the Witt class of (X, q), and hence gives rise to an additive surjection

$$\Delta\colon\ WQ(F) \to F/\wp F.$$

The kernel of Δ has been computed by H. Sah as follows. Let $I \subset W(F)$ be the fundamental ideal. Then

$$\operatorname{kernel}(\Delta) = I \cdot WQ(F).$$

The resulting additive isomorphism

$$WQ(F)/I \cdot WQ(F) \cong F/\wp F$$

should perhaps be regarded as an analogue of Pfister's isomorphism

$$I/I^2 \cong F^\bullet/F^{\bullet 2}.$$

(Compare Chapter III, § 5.2.)

Here is a simple example to illustrate the Arf invariant. Let (X, q) be a quadratic inner product space of rank 2. We continue to assume that the field F has characteristic 2. To any basis x, y with $(x|y)=1$ we associate the residue class

$$q(x)\,q(y) \quad \bigl(\operatorname{mod} \wp(F)\bigr).$$

This is an invariant of (X, q). For under an elementary change of basis $\bar{x} = x + \alpha y$ we have $q(\bar{x}) = q(x) + \alpha + \alpha^2 q(y)$, and therefore

$$q(\bar{x})\,q(y) = q(x)\,q(y) + \wp\bigl(\alpha q(y)\bigr).$$

From this formula we see that *the quadratic form q represents* 0 *if and only if* $q(x)\,q(y) \equiv 0 \bmod \wp(F)$. For if $q(x)\,q(y) \equiv 0$ and $q(y) \neq 0$, then we can choose α so that $q(\bar{x})\,q(y) = 0$. In fact we can actually choose a symplectic basis $\bar{x}$, $\bar{y}$ with $q(\bar{x}) = q(\bar{y}) = 0$, simply by setting $\bar{y} = \beta\bar{x} + y$ and choosing β appropriately.

If the field F is perfect, then the residue class $q(x)\,q(y) \bmod \wp(F)$ *is always a complete invariant for the quadratic inner product space.*

Proof. We may assume that $q(x) \neq 0$. Choosing an arbitrary representative $\Delta_0 = q(x)\,q(y) + \wp(\alpha)$ for the Arf invariant, the symplectic basis

$$\bar{x} = x/\sqrt{q(x)} \qquad \bar{y} = \alpha\bar{x} + y\sqrt{q(x)}$$

will satisfy

$$q(\bar{x}) = 1, \qquad q(\bar{y}) = \Delta_0.$$

Thus Δ_0 determines the isomorphism class of (X, q). ∎

In particular, *if F is a finite field of characteristic 2 it follows that there are precisely two isomorphism classes of quadratic inner product spaces of rank two over F.* For by inspecting the additive exact sequence $0 \to \mathbf{F}_2 \to F \xrightarrow{\wp} F$ we see that the cokernel $F/\wp(F)$ has order two.

Appendix 2. Hermitian Forms

Let R be an associative ring with 1, not necessarily commutative. By an *involution* of R (or more precisely an "involutory anti-automorphism") is meant an additive homomorphism $\alpha \mapsto \alpha^J$ from R to itself satisfying

$$(\alpha\beta)^J = \beta^J \alpha^J$$

and

$$(\alpha^J)^J = \alpha$$

for all α and β. Note that $1^J = 1$.

Examples. If R is commutative, then the identity map of R is an involution. For any multiplicative group Π, the integral group ring $\mathbf{Z}\Pi$ possesses a canonical involution which maps each group element σ to σ^{-1}. (Compare [Wall, § 5] as well as [Gel'fand-Mishchenko].) The ring of $n \times n$ matrices over a commutative ring has a canonical involution which maps each matrix to its transpose.

Let R be any fixed ring with involution, and let X be a left R-module. *Definition.* A *hermitian form* on X is a function

$$\varphi\colon X \times X \to R,$$

which is R-linear in the first variable and satisfies

$$\varphi(y, x) = \varphi(x, y)^J.$$

It follows that $\varphi(x, y)$ is bilinear over $\mathbf{Z}$, and that

$$\varphi(\alpha x, \beta y) = \alpha\, \varphi(x, y)\, \beta^J.$$

If the correspondence

$$y \mapsto \varphi(\ \ , y)$$

from X to $\mathrm{Hom}_R(X, R)$ is bijective, then φ is called a hermitian *inner product*, and the pair (X, φ) is called a *hermitian inner product module.*

Just as for symmetric inner product spaces and quadratic inner product spaces, one can define the concept of a *split* hermitian inner product space. Working modulo these split spaces, we obtain a *Witt group* of hermitian inner product spaces over R. The notation $W(R, J)$ will be used. In the commutative case, $W(R, J)$ has a natural ring structure. If J is the identity involution, then this coincides with the ordinary Witt ring $W(R)$.

If X is a free R-module with basis $e_1, \ldots, e_n$, then evidently any hermitian form φ on X is completely characterized by the matrix $[\varphi(e_i, e_k)]$, which is subject only to the requirement that

$$\varphi(e_k, e_i) = \varphi(e_i, e_k)^J.$$

The form φ is actually a hermitian inner product if and only if this matrix $[\varphi(e_i, e_k)]$ is invertible.

Now suppose that the ring R is commutative. Then the set

$$R_0 = \{\alpha \in \mathrm{R} \mid \alpha^J = \alpha\}$$

of fixed points forms a subring of R. The determinant of the matrix $[\varphi(e_i, e_k)]$ is evidently an element of R_0. If we choose some new basis $e'_1, \ldots, e'_n$ for X, the determinant will be multiplied by some arbitrary element in the image of the homomorphism

$$\text{norm}: R^\bullet \to R_0^\bullet$$

defined by $\text{norm}(\alpha) = \alpha\alpha^J$. *Definition.* The (multiplicative) residue class of $\det[\varphi(e_i, e_j)]$ modulo $\text{norm}(R^\bullet)$ is called the *determinant* of the hermitian space X. This determinant is a surprisingly powerful invariant. Compare Examples 2 and 4 below.

In the commutative case, every hermitian form over R gives rise to a quadratic form

$$Q(x) = \varphi(x, x)$$

over R_0. In particular, this function Q takes values in R_0 and the associated form

$$\begin{aligned}(x \mid y) &= Q(x+y) - Q(x) - Q(y)\\ &= \varphi(x, y) + \varphi(y, x)\end{aligned}$$

is bilinear over R_0.

Suppose now that F is a field with non-trivial involution. It follows from Galois theory that F is a quadratic Galois extension of the fixed field F_0.

Jacobson theorem. *Two hermitian inner product spaces over F are isomorphic if and only if their underlying quadratic spaces are isomorphic over F_0.*

In other words X is isomorphic to Y as hermitian space over F whenever there is an F_0-linear mapping from X onto Y preserving the quadratic function $Q(x) = \varphi(x, x)$. Thus the classification of hermitian inner product spaces over F is reduced to the classification of quadratic inner product spaces over F_0.

Proof by induction on the rank. First note that the field F contains an element α_0 with $\alpha_0+\alpha_0^J \neq 0$. In the characteristic 2 case, α_0 can be any element in the complement of F_0; while in the characteristic $\neq 2$ case we can take $\alpha_0=1$.

Next note that for any hermitian inner product space X over F, the associated bilinear form

$$(x|y)=\varphi(x,y)+\varphi(y,x)$$

is an inner product over F_0. In fact, given $x \neq 0$ we can certainly choose x' so that $\varphi(x,x') \neq 0$. After multiplying x' by an appropriate field element, we may assume that $\varphi(x,x')=\alpha_0$, and it follows that

$$(x|x')=\alpha_0+\alpha_0^J \neq 0.$$

Suppose now that the hermitian inner product spaces X and Y are isomorphic as F_0-quadratic spaces. Since the associated bilinear form $(x|x')$ is not identically zero, we can certainly choose a vector $x \in X$ and a corresponding vector $y \in Y$ with $Q(x)=Q(y) \neq 0$.

Since $\varphi(x,x) \neq 0$ it follows, just as in Chapter I, § 3, that the hermitian space X is isomorphic to an orthogonal sum $(Fx)\oplus(Fx)^{\perp}$. Similarly Y is isomorphic to $(Fy)\oplus(Fy)^{\perp}$. But evidently $(Fx)\cong(Fy)$.

Now let us pass to the underlying quadratic spaces over F_0, and apply the Witt cancellation theorem. (Compare Chapter I, §4, and Appendix 1.) It follows that $(Fx)^{\perp}$ is isomorphic as quadratic space to $(Fy)^{\perp}$. Using the induction hypothesis, it follows that $(Fx)^{\perp}\cong(Fy)^{\perp}$ as hermitian space, and therefore $X\cong Y$. □

Corollary. *With $F \supset F_0$ as above, there is an exact sequence*

$$0 \to W(F,J) \to WQ(F_0) \to WQ(F)$$

of $W(F_0)$-modules.

Proof. Given a non-zero element of $W(F,J)$, we can clearly choose a representative hermitian inner product space X which is anisotropic:

$$\varphi(x,x) \neq 0 \quad \text{for } x \neq 0.$$

It follows that the underlying quadratic space is also anisotropic, and hence represents a non-zero element of $WQ(F_0)$. Thus the natural homomorphism $W(F,J) \to WQ(F_0)$ is injective.

Next let us look at the composition $W(F,J) \to WQ(F_0) \to WQ(F)$. Choose a basis $\{1,\alpha\}$ for F over F_0. First consider a hermitian space X of dimension 1 over F. Then X has a basis vector e_1 with $\varphi(e_1,e_1) \in F_0^{\bullet}$. It follows that X is 2-dimensional over F_0 with basis $\{e_1,e_2\}$ where $e_2=\alpha e_1$, and

$$Q(e_2)=\alpha\alpha^J Q(e_1), \quad (e_1|e_2)=(\alpha+\alpha^J)Q(e_1).$$

(It may be convenient to choose α so that $\alpha+\alpha^J=1$ in the characteristic 2 case, or $\alpha+\alpha^J=0$ in the characteristic $\neq 2$ case; but this is not necessary.)

Now let us pass to the induced quadratic space $F \underset{F_0}{\otimes} X$ of dimension 2 over F. Consider the non-zero vector $\alpha\otimes e_1 - 1\otimes e_2$ in this induced space. Evidently

$$Q(\alpha\otimes e_1 - 1\otimes e_2) = \alpha^2 Q(e_1) + Q(e_2) - \alpha(e_1|e_2)$$
$$= (\alpha^2 + \alpha\alpha^J - \alpha(\alpha+\alpha^J))\, Q(e_1) = 0.$$

Therefore this induced quadratic space is split. Since any hermitian inner product space over F is clearly an orthogonal sum of 1-dimensional spaces, this proves that the composition $W(F,J)\to WQ(F_0)\to WQ(F)$ is zero.

Conversely let Y be a quadratic space over F_0 which maps to zero in $WQ(F)$. After eliminating any split orthogonal summand, we may assume that Y is anisotropic. Since $F\underset{F_0}{\otimes} Y$ is split, it certainly contains a vector $y\neq 0$ with $Q(y)=0$. Setting

$$y = \alpha\otimes y_1 - 1\otimes y_2$$

with $y_1, y_2\in Y$, we have

$$Q(y) = \alpha^2 Q(y_1) + Q(y_2) - \alpha(y_1|y_2) = 0.$$

Substituting

$$\alpha^2 = \alpha(\alpha+\alpha^J) - \alpha\alpha^J$$

and recalling that 1 and α are linearly independent over F_0, it follows that

$$(y_1|y_2) = (\alpha+\alpha^J)\, Q(y_1)$$
$$Q(y_2) = \alpha\alpha^J\, Q(y_1),$$

where $Q(y_1)\neq 0$ since Y is anisotropic. This proves that the subspace of Y spanned by y_1 and y_2 is isomorphic to the underlying quadratic space of a hermitian space over F spanned by a vector y_1 with $\varphi(y_1,y_1)=Q(y_1)$, where αy_1 corresponds to the vector y_2.

Now express Y as an orthogonal sum $(F_0 y_1 \oplus F_0 y_2)\oplus(F_0 y_1\oplus F_0 y_2)^\perp$. The second summand has smaller rank, is also anisotropic, and also represents an element of the kernel. A straightforward induction now completes the proof. □

Here are some examples worked out in more detail. (Compare [Milnor, 1969].) *In each case we assume that the involution J is not the identity.*

Example 1. If F is a finite field, then a hermitian inner product space splits if and only if it has even rank. The rank is a complete invariant; and $W(F,J)\cong \mathbf{Z}/2$.

Note that the description is exactly the same whether the characteristic is 2 or $\neq 2$. The proof will be left to the reader.

Example 2. If F is a local field, or a function field in one variable over a finite field, then the rank and determinant of a hermitian inner product space form a complete system of invariants. The kernel of the rank homomorphism $W(F, J) \to \mathbf{Z}/2$ is an ideal, additively isomorphic to $F_0^\bullet/\text{norm}\, F^\bullet$, and with square equal to 0.

Again the characteristic 2 case is not distinguished in any way. The proof can be sketched as follows. It suffices to note that any space X of rank $\geqq 2$ over F has rank $\geqq 4$ over F_0, hence the quadratic equation $Q(e_1)=1$ has a solution. (Using the Hasse-Minkowski theorem in the global case; compare Chapter II, § 3. For the characteristic 2 case, see [Arf, pp. 164–167].) Hence X is isomorphic to an orthogonal sum $(F e_1) \oplus (F e_1)^\perp$. Continuing inductively, we find an orthogonal basis $e_1, \ldots, e_n$ with $\varphi(e_i, e_i) = Q(e_i) = 1$ for $i < n$. Hence the determinant $\varphi(e_n, e_n)$ determines the structure of X completely.

Example 3. If F is the field $\mathbf{C}$ of complex numbers, with complex conjugation as involution, then clearly the rank and signature of the underlying quadratic space form a complete system of invariants. Hence $W(\mathbf{C}, \text{conjugation}) \cong \mathbf{Z}$.

Example 4. If F is an algebraic extension of the rationals, then every conjugation preserving embedding ω: $F \to \mathbf{C}$ gives rise to a signature homomorphism

$$\omega_*: W(F, J) \to W(\mathbf{C}, \text{conjugation}) \cong \mathbf{Z}.$$

The rank, determinant, and these various signatures form a complete system of invariants for a hermitian inner product space over F. For details, the reader is referred to [Landherr].

To conclude this section, we remark that Jacobson's theorem applies also in certain non-commutative situations. For the statement to make sense however, we must assume that the fixed point set R_0 of the involution J is a subring, contained in the center of R. This is notably the case if R is a *quaternion algebra* with basis 1, i, j, and $ij = -ji$ over R_0, where i^2 and j^2 belong to $R_0^\bullet$. Here R_0 should be a field of characteristic $\neq 2$. The involution is defined by $i^J = -i$, $j^J = -j$. Note that the norm $\xi \xi^J$ of an element $\xi = \alpha + \beta i + \gamma j + \delta ij$ is equal to

$$\alpha^2 - \beta^2 i^2 - \gamma^2 j^2 + \delta^2 i^2 j^2.$$

If $\xi \xi^J \neq 0$ for $\xi \neq 0$, or in other words if the associated inner product space

$$\langle 1 \rangle \oplus \langle -i^2 \rangle \oplus \langle -j^2 \rangle \oplus \langle i^2 j^2 \rangle$$

over R_0 is anisotropic, then clearly R is a division algebra (= skew field). In this case, Jacobson's argument applies just as before. There is a canonical embedding

$$0 \to W(R, J) \to WQ(R_0),$$

and two hermitian inner product spaces are isomorphic if and only if their underlying quadratic spaces are isomorphic over R_0.

Jacobson remarks also that the "determinant" of a hermitian space can still be defined, in this non-abelian context, as an element of $R_0^\bullet/\text{norm}(R^\bullet)$. The definition is based on work of E.H. Moore. (Compare [Dyson].)

Appendix 3. The Hasse-Minkowski Theorem

The Hasse-Minkowski theorem is one of the most beautiful results in algebraic number theory. The proof which follows assumes some knowledge of Class Field Theory, as described in [Lang] or [Cassels-Fröhlich]. In the case of the rational field, it is possible to give a more elementary proof. (See [Serre] or [Borevich-Shafarevich].) For a complete and self-contained proof in the general case, see [O'Meara].

First some definitions. Let X be an inner product space over a field F. Then X is said to *represent* a field element α if there exists a non-zero vector $x \in X$ with $x \cdot x = \alpha$.

Henceforth we assume that F has characteristic $\neq 2$.

Lemma 1. *If a space X over F represents* 0, *then it represents every element of F.*

For if X represents 0 then X admits a hyperbolic plane as direct summand (Chapter I, § 6), and a hyperbolic plane clearly represents all field elements. ∎

Corollary. *A space X represents the element $\alpha \neq 0$ if and only if the orthogonal sum $X \oplus \langle -\alpha \rangle$ represents* 0.

(In other words an inhomogeneous equation in n variables can be expressed as a homogeneous equation in $n+1$ variables.) The proof is immediate. ∎

Now suppose that F is a *global field*. That is, F is either finite over the rational numbers, or finitely generated of transcendence degree 1 over a finite field. We continue to assume that F has characteristic $\neq 2$.

For every (non-trivial) valuation v of F, let F_v denote the completion, and let X_v denote the induced inner product space $F_v \otimes X$ over F_v. Let α be some fixed element of F.

Hasse-Minkowski theorem. *The inner product space X represents α if and only if X_v represents α for every (non-trivial) valuation v of F.*

Both archimedean and non-archimedean valuations v must be included.

In order to prove this theorem, it will be convenient to break it up into two parts, according as the field element α is zero or non-zero:

Assertion A_n. Let $\alpha \in F^\bullet$. *A space X of rank n over F represents α if and only if X_v represents α for every v.*

Using the corollary to Lemma 1 above, this is completely equivalent to the following statement.

Assertion A'_n. *A space Y of rank $n+1$ over F represents* 0 *if and only if the completion Y_v represents* 0 *for every valuation v.*

Note the shift from dimension n to $n+1$ in Assertion A'_n.

In order to prove these two statements, we will pass back and forth between the two forms, first proving A_2, and then showing that

$$A'_2 \Rightarrow A_3 \Rightarrow A'_n$$

for $n \geqq 4$. The proof of A_1 will be given last, since it is completely irrelevant to the rest of the argument.

Proof of A_2. Suppose that $X \cong \langle u_1 \rangle \oplus \langle u_2 \rangle$. We must try to solve the equation

$$u_1 \xi^2 + u_2 \eta^2 = \alpha,$$

where $\alpha \neq 0$. Alternatively, setting $u = -u_2/u_1$ and $\beta = \alpha/u_1$, this can be written as

$$\xi^2 - u\eta^2 = \beta. \tag{1}$$

Let K denote the extension field $F(\sqrt{u}) = F(\sqrt{-u_2/u_1})$. Then the Eq. (1) possesses a solution $\xi, \eta \in F$ if and only if β belongs to the image of the norm homomorphism

$$\mathrm{norm}_{K/F}\colon K^\bullet \to F^\bullet.$$

If $K \neq F$ this is clear, since $\mathrm{norm}(\xi + \eta\sqrt{u}) = \xi^2 - \eta^2 u$. If $K = F$ it is clear since $u_2 \in -u_1 F^{\bullet 2}$, so the inner product space X splits and the Eq. (1) always has a solution.

Now recall the following:

Hasse norm theorem. *Let K be a cyclic Galois extension of the global field F. Then an element α of $F^\bullet$ belongs to the image of the homomorphism* $\mathrm{norm} = \mathrm{norm}_{K/F}\colon K^\bullet \to F^\bullet$ *if and only if α belongs to the image of*

$$\mathrm{norm}\colon K^\bullet_w \to F^\bullet_{w|F}$$

for every valuation w of K.

This is proved in [Lang, p. 195] or [Cassels-Fröhlich, p. 185].

If the Eq. (1) has a solution in F_v for every v, then β is a norm from $K_w^\bullet$ for every w, hence β is a norm from K by the Hasse norm theorem. This completes the proof of Assertion A_2.

For the proof of A_3, the following will be needed.

Lemma 2. *Let X be an inner product space of rank 3 and determinant d over a field F of characteristic $\neq 2$. To any field element $\alpha \neq 0$ we associate the extension field $K = F(\sqrt{-\alpha d})$. Then X represents α if and only if the induced inner product space $K \underset{F}{\otimes} X$ over K represents 0.*

Proof. If X represents α, then $X \cong \langle\alpha\rangle \oplus \langle\beta\rangle \oplus \langle\gamma\rangle$ for some elements β, γ, where

$$d \in \alpha\beta\gamma F^{\bullet 2}.$$

Over the extension field $K = F(\sqrt{-\alpha d}) = F(\sqrt{-\beta\gamma})$, we see that γ is equal to $-\beta$ multiplied by a square. Therefore the space $K \otimes (\langle\beta\rangle \oplus \langle\gamma\rangle)$ splits, hence $K \otimes X$ represents 0.

Conversely suppose that $K \otimes X$ represents 0. We may suppose that K is a proper extension of F, since if $K = F$ then X itself represents 0, hence X represents α. Thus there exist two vectors $x, y \in X$, not both zero, so that

$$(x + \sqrt{-\alpha d}\, y) \cdot (x + \sqrt{-\alpha d}\, y) = 0.$$

In other words

$$x \cdot x - \alpha d\, y \cdot y = 0, \qquad x \cdot y = 0.$$

We may assume that the inner product $x \cdot x = \alpha d\, y \cdot y$ is not 0, since otherwise X would represent 0. Hence the two orthogonal vectors x and y form part of an orthogonal base x, y, z, with

$$X \cong \langle x \cdot x\rangle \oplus \langle y \cdot y\rangle \oplus \langle z \cdot z\rangle,$$

and

$$d \in (x \cdot x)(y \cdot y)(z \cdot z) F^{\bullet 2}.$$

Substituting the equation $x \cdot x = \alpha d\, y \cdot y$, we see that $z \cdot z$ is equal to α multiplied by an element of $F^{\bullet 2}$. This proves Lemma 2. □

Proof that A_2' implies A_3. Fixing the space X of rank 3, and the field element $\alpha \neq 0$, let $K = F(\sqrt{-\alpha d})$ as in the lemma. For each valuation w of K we clearly have

$$K_w = F_v(\sqrt{-\alpha d})$$

where $v = w|F$. Thus if X_v represents α for every v, then $(K \otimes X)_w$ represents 0 for every w. Using the Assertion A_2', which has already been proved, it follows that $K \otimes X$ represents 0, hence X represents α.

Proof that A_3 implies A_n' for $n \geqq 4$. Let X be a space of rank $n+1 \geqq 5$. Then X is isomorphic to a sum $Y \oplus Z$ where $Z \cong \langle u_1\rangle \oplus \langle u_2\rangle \oplus \langle u_3\rangle$ has rank 3, and $Y \cong \langle u_1'\rangle \oplus \cdots \oplus \langle u_{n-2}'\rangle$ has rank $\geqq 2$.

Let $T = T(u_1, u_2, u_3)$ denote the finite set consisting of all (equivalence classes of) valuations v such that either

(1) v is archimedean or dyadic, or

(2) $|u_1|_v \neq 1$ or $|u_2|_v \neq 1$ or $|u_3|_v \neq 1$.

For $v \notin T$ we see as in Chapter II, § 3.4 that the completed space Z_v necessarily represents 0.

Suppose now that X_v represents 0. Then we can certainly choose vectors $y_v \in Y_v$ and $z_v \in Z_v$, not both zero, so that $y_v \cdot y_v + z_v \cdot z_v = 0$. *In fact these vectors can be chosen so that*

$$y_v \cdot y_v = -z_v \cdot z_v \neq 0.$$

For if our first choice of y_v and z_v yields $y_v \cdot y_v = z_v \cdot z_v = 0$, then either Y represents 0, in which case we can choose an arbitrary z'_v with $z'_v \cdot z'_v \neq 0$ and apply Lemma 1, or Z represents 0 in which case any $y'_v \cdot y'_v \neq 0$ will do.

We will also make use of the following.

Weak approximation theorem. *Given finitely many inequivalent valuations $v_1, \ldots, v_t$ on a field F, the image of the diagonal embedding*

$$F \to F_{v_1} \times \cdots \times F_{v_t}$$

is everywhere dense.

This is proved for example in Lang, *Algebra*, Addison-Wesley 1965, p. 285.

Now consider the $(n-2)$-dimensional vector space Y over F, and the set $T = \{v_1, \ldots, v_t\}$. Applying this approximation theorem to each of the $n-2$ coordinates, we see that the image of the diagonal embedding $Y \to Y_{v_1} \times \cdots \times Y_{v_t}$ is dense. In particular, we can choose an element $y \in Y$ which is so close to y_v for each $v \in T$ that the ratio $(y \cdot y)/(y_v \cdot y_v)$ is a square in $F_v^\bullet$.

Let us apply Assertion A_3 to the space Z. For each $v \in T$ the completion Z_v represents $-y_v \cdot y_v$ and therefore represents $-y \cdot y$. But for each $v \notin T$ the completion represents 0, and therefore represents $-y \cdot y$. Applying A_3, it follows that Z itself represents $-y \cdot y$, and therefore $X = Y \oplus Z$ represents 0. This completes the proof of A_n and A'_n for $n \geqq 2$.

To conclude the proof of the Hasse-Minkowski theorem, we must prove Assertion A_1. Given $X \cong \langle u \rangle$, and given $\alpha \neq 0$ in F, we must solve the equation

$$u \xi^2 = \alpha.$$

Setting $\alpha/u = \beta$, this can be written as $\xi^2 = \beta$. Thus we must prove the following.

Square theorem. *If the field element $\beta \in F^{\bullet}$ is a square in the completion F_v for every v, then β is a square in F.*

This theorem follows easily from the basic inequalities of global class field theory. Recall that the *idele group* $A_F^{\bullet}$ is the group of units in the ring $A_F \subset \prod_v F_v$ consisting of all elements (a_v) in the cartesian product which satisfy the condition $|a_v|_v \leqq 1$ for almost all v. (In forming this cartesian product, one of course chooses just one valuation v in each non-trivial equivalence class of valuations.) The quotient $A_F^{\bullet}/F^{\bullet}$ is called the *idele class group* C_F. For any finite extension $K \supset F$, the local norm homomorphisms $K_w^{\bullet} \to F_{w|F}^{\bullet}$ combine to yield the global norm homorphisms $A_K^{\bullet} \to A_F^{\bullet}$, and $C_K \to C_F$. If K is cyclic of degree m over F, then the inequalities of class field theory state that *the index of the subgroup* $\operatorname{norm}_{K/F} C_K \subset C_F$ *is equal to* m. See [Lang, p. 192] or [Cassels-Fröhlich, p. 179].

Proof of the square theorem. Given $\beta \in F^{\bullet}$, let $K = F(\sqrt{\beta})$. If β is a square in F_v for every v, then $K_w = F_{w|F}$ for every valuation w of K, so the norm homomorphism $A_K^{\bullet} \to A_F^{\bullet}$ is surjective. Therefore the norm homomorphism $C_K \to C_F$ is surjective, and the degree m must be 1. Thus $\sqrt{\beta} \in F$, which completes the proof of the square theorem and the Hasse-Minkowski theorem. □

Now consider a more general situation. Let X and Y be two inner product spaces over a field F with $\operatorname{rank}(X) \geqq \operatorname{rank}(Y)$.

Definition. The space X is said to *represent* Y if $X \cong Y \oplus Z$ for some Z.

If Y has rank 1, say $Y \cong \langle u \rangle$, then clearly X represents Y if and only if X represents u.

Corollary 1. *Suppose again that F is a global field. If the completion X_v represents Y_v for every valuation v, then X represents Y.*

Proof. This is certainly true if Y has rank 1. If the rank of Y is greater than 1, then, setting $Y = Y' \oplus \langle u \rangle$, we may assume inductively that X represents Y', say

$$X \cong Y' \oplus Z'. \tag{2}$$

By hypothesis, for each v there exists a space $Z(v)$ over F_v so that

$$\begin{aligned} X_v &\cong Y_v \oplus Z(v) \\ &\cong Y'_v \oplus \langle u \rangle_v \oplus Z(v). \end{aligned}$$

Comparing this isomorphism with the completion of (2), and applying the Witt cancellation theorem, we obtain

$$Z'_v \cong \langle u \rangle_v \oplus Z(v).$$

Thus Z'_v represents u for every v. By the Hasse-Minkowski theorem, it follows that Z' represents u, say

$$Z' \cong \langle u \rangle \oplus Z''.$$

Together with (2), this completes the proof. □

Corollary 2. *Two spaces X and Y over F are isomorphic if and only if X_v is isomorphic to Y_v for every v. In particular X splits if and only if X_v splits for every v.*

Proof. This is just the special case $\operatorname{rank}(X) = \operatorname{rank}(Y)$ of Corollary 1, using Chapter I, § 6.3. □

Here is still another formulation of the Hasse-Minkowski theorem.

Corollary 3. *Consider a quadratic equation $\sum \alpha_{ij} \xi_i \xi_j + \sum \beta_k \xi_k + \gamma = 0$, in n variables, with coefficients in the global field F. If this equation has a solution in F_v for every v, then it has a solution in F.*

Note that the corresponding statement for equations of higher degree, or for systems of quadratic equations, would be false. Here is a trivial example. The 6-th degree equation

$$(\xi^2+1)(\xi^2+17)(\xi^2-17)=0$$

has a solution in $\mathbf{Q}_v$ for every v, but has no rational solution. Similarly, consider the simultaneous quadratic equations

$$\begin{aligned} \xi^2+\eta&=0, \\ (\eta-\zeta)(\eta-\zeta+16)&=0, \\ \zeta^2&=17^2. \end{aligned}$$

For any solution (ξ, η, ζ) we must have $\zeta = \pm 17$ hence $\eta = 1$, 17, -17 or -33, and $\xi^2+\eta=0$. Again there is a solution in $\mathbf{Q}_v$ for every v, but no solution in $\mathbf{Q}$.

Proof of Corollary 3. Write the given equation as

$$\alpha(x,x)+\beta(x)+\gamma=0, \tag{3}$$

where α is a symmetric bilinear form on the vector space $X = F^n$, and where β is an element of the dual vector space $\operatorname{Hom}(X, F)$. Let N be the null space of the linear mapping $x \mapsto \alpha(x, \ )$ from X to $\operatorname{Hom}(X, F)$. Counting dimensions, we see that the sequence

$$0 \to N \to X \to \operatorname{Hom}(X,F) \to \operatorname{Hom}(N,F) \to 0$$

is exact.

If the element $\beta \in \mathrm{Hom}(X, F)$ restricts to a non-zero element of $\mathrm{Hom}(N, F)$, then it is easy to choose $x \in N$ so as to satisfy the required equation

$$0+\beta(x)+\gamma=0. \tag{3'}$$

Suppose then that β maps to zero in $\mathrm{Hom}(N, F)$. Then β lifts to an element of X, say

$$\beta(x)=2\alpha(x_0, x)$$

for every $x \in X$. The substitution $x=y-x_0$ now reduces Eq. (3) to the form

$$\alpha(y, y)=\gamma', \tag{4}$$

with $\gamma'=\beta(x_0)-\alpha(x_0, x_0)-\gamma$. If $N=0$, so that X is an inner product space, we can apply the Hasse-Minkowski theorem to Eq. (4) and obtain a solution. If $N \neq 0$, we must simply choose a complementary direct summand

$$X=N \oplus X'.$$

Then α restricted to X' is an inner product, so the Hasse-Minkowski theorem applies, and Eq. (4) has a solution $y \in X' \subset X$. This completes the proof of Corollary 3. $\square$

Appendix 4. Gauss Sums, the Signature mod 8, and Quadratic Reciprocity

Let L be a free $\mathbf{Z}$-module of rank n, provided with a $\mathbf{Z}$-valued symmetric bilinear form $x \cdot y$ with non-zero determinant. We denote the signature of this form by σ, and the absolute value of the determinant by d. An expression for $\exp(2\pi i \sigma/8)$ as a finite exponential sum was given by H. Braun in 1940. (Explicitly, she showed that

$$(2a)^{n/2}\sqrt{d}\exp(2\pi i \sigma/8) = \sum_{x \in L/aL} \exp(2\pi i\, x \cdot x/a)$$

where $a = 8d^3$. It follows that σ mod 8 is determined by the a^n numbers $x \cdot x$ modulo a.) We will describe a closely related formula which has recently been obtained by J. Milgram. (Compare the discussion in Chapter II, § 5.)

As in Chapter II, we will say that L is of *type* II if the congruence

$$(1) \qquad x \cdot x \equiv 0 \mod 2$$

is satisfied for every $x \in L$. Let $L^\#$ denote the dual lattice, consisting of all $u \in \mathbf{Q} \otimes L$ satisfying the condition $u \cdot L \subset \mathbf{Z}$. Then the quotient $L^\#/L$ is a finite abelian group of order d. If L is of type II, then setting

$$\varphi(u) = \tfrac{1}{2} u \cdot u \qquad \text{modulo } \mathbf{Z},$$

we obtain a well defined quadratic function $\varphi\colon\ L^\#/L \to \mathbf{Q}/\mathbf{Z}$.

Theorem (Milgram). *If L is of type II, then the Gauss sum*

$$\sum_{u \in L^\#/L} \exp(2\pi i\, \varphi(u))$$

is defined and is equal to $\sqrt{d}\exp(2\pi i \sigma/8)$.

The original proof of this formula was a rather delicate argument involving the Poisson summation formula. The following proof, suggested by Knebusch, is quite a bit easier.

Consider lattices L of type II in a fixed rational inner product space. We will denote the d-fold sum $\sum_{L^\#/L} \exp(2\pi i\, \varphi(u)) = \sum_{L^\#/L} \exp(\pi i\, u \cdot u)$ briefly by the symbol $G(L)$.

Lemma 1. *If $L_1 \subset L$ is a sub-lattice of index k, then $G(L_1) = kG(L)$.*

Proof. Evidently $L_1 \subset L \subset L^\# \subset L_1^\#$ where the index of each lattice in the next is equal to k, or $d(L)$, or k respectively. Let $x_1, \ldots, x_{kd(L)}$ be a complete set of coset representatives for $L_1^\#$ modulo L. Then the Gauss sum $G(L_1)$ can be written as

$$G(L_1) = \sum_{x_j} \sum_{u \in L/L_1} \exp(2\pi i \varphi(x_j + u)).$$

If we substitute

$$\varphi(x_j + u) \equiv \varphi(x_j) + x_j \cdot u \qquad (\mathrm{mod}\ \mathbf{Z}),$$

this becomes

$$G(L_1) = \sum_{x_j} \exp(2\pi i \varphi(x_j)) \sum_{u \in L/L_1} \exp(2\pi i (x_j \cdot u)).$$

But for each fixed x_j the k-fold sum

$$\sum_{u \in L/L_1} \exp(2\pi i (x_j \cdot u)) \tag{2}$$

can be evaluated as follows. If x_j happens to belong to $L^\#$, this sum is evidently equal to $1 + \cdots + 1 = k$. If $x_j \notin L^\#$ then the correspondence

$$u \mapsto \exp(2\pi i (x_j \cdot u))$$

defines a non-trivial homomorphism from L/L_1 to $\mathbf{C}^\bullet$, so a standard argument [Lang, p. 82] shows that the sum (2) is zero. Therefore $G(L_1)$ is equal to

$$\sum_{x_j \in L^\#/L} \exp(2\pi i \varphi(x_k))\, k = kG(L)$$

as asserted. □

Evidently $d(L_1) = k^2 d(L)$; so it follows from Lemma 1 that

$$G(L)/\sqrt{d(L)} = G(L_1)/\sqrt{d(L_1)}.$$

In fact the complex number $G(L)/\sqrt{d(L)}$ is completely independent of the lattice L, and depends only on the ambient rational inner product space. This is clear, since any two lattices L and L' spanning the same rational space must contain a common sub-lattice $L \cap L'$ which has finite index in each of them.

To evaluate this invariant $G(L)/\sqrt{d(L)}$ of the inner product space $\mathbf{Q} \otimes L$, we recall that every rational inner product space is isomorphic to an orthogonal sum of 1-dimensional spaces, and hence contains a lattice of type II which splits as an orthogonal sum of 1-dimensional lattices. Note that the invariant $G(L)/\sqrt{d(L)}$ is multiplicative with respect to orthogonal sums. For the identity

$$G(L_1 \oplus L_2) = G(L_1)\, G(L_2)$$

is easily verified, and the identity

$$d(L_1 \oplus L_2) = d(L_1)\, d(L_2)$$

is familiar. Thus to compute this invariant $G(L)/\sqrt{d(L)}$ for any rational inner product space, it suffices to compute it for a 1-dimensional inner product space.

The following elementary observation will be needed for the computation in the 1-dimensional case.

Lemma 2. *For any constant* $c > 0$, *the integral* $\int_0^A \exp(c \pi i s^2)\, ds$ *tends to a well defined finite limit as* $A \to \infty$.

To prove that the imaginary part of this integral converges, substitute $u = c s^2$ and integrate between successive integer values of u, noting that the terms of the resulting series alternate in sign, with absolute values tending monotonely to zero. Convergence of the real part is proved similarly, using half-integer values. ☐

Consider now a 1-dimensional lattice of type II, say $L \cong \langle 2m \rangle$. Suppose, to fix our ideas, that $m > 0$. Evidently $L^{\#}/L$ is cyclic of order $2m$, and $G(L)$ is equal to the $2m$-fold sum

$$\sum_{k=0}^{2m-1} \exp(\pi\, i\, k^2/2m).$$

To evaluate this sum, following Dirichlet and Landau, we introduce an associated periodic function $f: \mathbf{R} \to \mathbf{C}$ of period 1, where

$$f(t) = \sum_{k=0}^{2m-1} \exp\bigl(\pi\, i (k+t)^2/2m\bigr)$$

for $0 \leqq t \leqq 1$. Thus $f(0) = f(1)$ is equal to the Gauss sum $G(L)$. (Compare [Lang, p. 88].)

Since f is continuous and piecewise smooth, its Fourier series expansion converges to f everywhere. (See, for example, Titchmarsh, Theory of Functions; or Courant and Hilbert, Volume 1.) We will write this Fourier series in the form

$$f(t) = \sum_{-\infty}^{\infty} a_n \exp(-2\pi i n t),$$

where

$$a_n = \int_0^1 f(t) \exp(2\pi i n t)\, dt.$$

Note in particular that the Gauss sum $G(L) = f(0)$ is equal to $\sum_{-\infty}^{\infty} a_n$.

To evaluate the coefficient a_n, first substitute the definition of $f(t)$, obtaining

$$a_n = \sum_{k=0}^{2m-1} \int_0^1 \exp\left(2\pi i \left(\frac{(k+t)^2}{4m} + nt\right)\right) dt.$$

Next complete the square, so as to obtain the congruence

$$\frac{(k+t)^2}{4m} + nt \equiv (k+t+2mn)^2/4m \qquad (\operatorname{mod} \mathbf{Z}),$$

and then substitute $s = k+t+2mn$. This yields

$$\begin{aligned} a_n &= \sum_{k=0}^{2m-1} \int_{k+2mn}^{k+1+2mn} \exp(2\pi i s^2/4m)\, ds \\ &= \int_{2mn}^{2m(n+1)} \exp(\pi i s^2/2m)\, ds. \end{aligned}$$

Now sum over n, so as to obtain the formula

$$G(L) = \sum a_n = \int_{-\infty}^{\infty} \exp(\pi i s^2/2m)\, ds,$$

where the improper integral is well defined by Lemma 2. In fact, substituting $u = s/\sqrt{2m}$, it follows that $G(L) = G(\langle 2m \rangle)$ is equal to

$$\sqrt{2m} \int_{-\infty}^{\infty} \exp(\pi i u^2)\, du.$$

Thus the ratio

$$G(\langle 2m \rangle)/\sqrt{2m} = \int_{-\infty}^{\infty} \exp(\pi i u^2)\, du$$

is independent of m. To evaluate this integral, we simply evaluate the Gauss sum for the case $m=1$, obtaining the identity

$$G(\langle 2m \rangle)/\sqrt{2m} = (1+i)/\sqrt{2} = \exp(2\pi i/8).$$

Similarly $G(\langle -2m \rangle)/\sqrt{2m}$ is equal to the complex conjugate $\exp(-2\pi i/8)$. Thus we have shown that the invariant $G(L)/\sqrt{d(L)}$ is equal to $\exp(2\pi i \sigma/8)$ for every 1-dimensional lattice L. The corresponding formula for an orthogonal sum of 1-dimensional lattices, and hence for an arbitrary lattice, now follows immediately. This completes the proof of Milgram's theorem. □

The formula of Braun can be recovered from that of Milgram as follows. Let L be any lattice in a rational inner product space, subject only to the hypothesis that $x \cdot y \in \mathbf{Z}$ for $x, y \in L$. As before we set $d = |\text{determinant}| > 0$, and $n = rk(L)$.

Corollary. *If q is a multiple of $2d$, then*

$$\sum_{x\in L/qL} \exp(\pi i x\cdot x/q)=q^{n/2}\sqrt{d}\exp(2\pi i\sigma/8).$$

Evidently the formula quoted at the beginning of this Appendix is an immediate consequence, taking $q=a/2$.

Proof of the corollary. Consider the new inner product $q x\cdot y$ on the lattice $L^{\#}$. Note that $L^{\#}$, with this new inner product, is of type II, and has dual lattice equal to $q^{-1}L$. Applying Milgram's theorem we obtain

$$\sum_{u\in q^{-1}L/L^{\#}} \exp(\pi i q u\cdot u)=\sqrt{q^n/d}\exp(2\pi i\sigma/8).$$

Now substitute $u=q^{-1}x$ and multiply both sides of this equation by d. The left hand side becomes

$$d\sum_{x\in L/qL^{\#}} \exp(\pi i x\cdot x/q)=\sum_{x\in L/qL} \exp(\pi i x\cdot x/q),$$

since $qL^{\#}$ contains qL as a subgroup of index d. This completes the proof. ∎

Knebusch points out that Milgram's formula is closely related to a version of the quadratic reciprocity law due to [Weil, 1964]. Given L as above, let $(L^{\#}/L)_p$ denote the p-primary component of the finite abelian group $L^{\#}/L$. Then $L^{\#}/L$ decomposes as the orthogonal sum of the various $(L^{\#}/L)_p$, hence the Gauss sum $G(L)$ can be expressed correspondingly as a product $\prod G((L^{\#}/L)_p)$, where all but finitely many of the factors are equal to 1. Similarly the order d of $L^{\#}/L$ splits as the product of its p-primary components d_p.

Lemma 3. *The ratio $\gamma_p(L)=G((L^{\#}/L)_p)/\sqrt{d_p}$ depends only on the p-adic completion $\mathbf{Q}_p\otimes L$ of the inner product space $\mathbf{Q}\otimes L$. The correspondence $\mathbf{Q}_p\otimes L\mapsto\gamma_p(L)$ gives rise to a homomorphism from the finite additive group $W(\mathbf{Q}_p)$ to the multiplicative group of roots of unity in $\mathbf{C}^{\bullet}$.*

Briefly we say that γ_p is a *character* of the Witt group $W(\mathbf{Q}_p)$.

The proof of the first statement is completely analogous to the proof of Lemma 1. One simply uses the p-adic integers $\mathbf{Z}_p$ and the p-adic field $\mathbf{Q}_p$ in place of $\mathbf{Z}$ and $\mathbf{Q}$. To prove the second statement, suppose that the completion $\mathbf{Q}_p\otimes L$ is a split inner product space. Then this completion has inner product matrix $\begin{pmatrix}0 & I\\ I & 0\end{pmatrix}$ with respect to a suitable basis, and this basis spans a $\mathbf{Z}_p$-lattice which is self-dual. But the existence of such a self-dual lattice implies that $\gamma_p(L)=1$. Since the function γ_p is clearly multiplicative with respect to orthogonal sums, this proves the lemma. ∎

Let us define $\gamma_\infty(L)$ to be the root of unity $\exp(-2\pi i\,\sigma/8)$. Evidently this depends only on the real completion $\mathbf{R}\otimes L$ of the inner product space $\mathbf{Q}\otimes L$.

Weil reciprocity theorem. *For any lattice L, the product* $\prod_{p\leqq\infty}\gamma_p(\mathbf{Q}\otimes L)$ *is equal to* 1.

Proof. This follows immediately from Milgram's theorem. $\square$

Let us see what this reciprocity formula means in the rank 1 case. Suppose then that the lattice L is spanned by a single vector l_1. Suppose also, to fix our ideas, that $l_1\cdot l_1=4m$ with m odd. We will write briefly $L=\langle 4m\rangle$.

Lemma 4. *The character* $\gamma_2\langle 4m\rangle$ *is equal to* $\exp(2\pi i\,m/8)$.

In fact $L^\#/L$ is cyclic of order $|4m|$, generated by $l_1/4m$. Hence the 2-primary component $(L^\#/L)_2$ is cyclic of order 4, generated by $l_1/4$, with $\varphi(l_1/4)\equiv m/8\,(\text{mod}\ \mathbf{Z})$. It follows easily that

$$\begin{aligned}\gamma_2\langle 4m\rangle &= \sum_{j=1}^{4}\exp(2\pi i\,j^2\,m/8)/\sqrt{4}\\ &=\exp(2\pi i\,m/8).\quad\square\end{aligned}$$

Next suppose that m is an odd prime p.

Lemma 5. *The character* $\gamma_p\langle 4p\rangle$ *is equal to* $\exp(2\pi i(1-p)/8)$.

Proof. This follows from Lemma 4, by solving the reciprocity equation

$$\gamma_2\langle 4p\rangle\,\gamma_p\langle 4p\rangle\,\gamma_\infty\langle 4p\rangle=1.\quad\square$$

More generally suppose that $m=p\,u$, where u is relatively prime to p.

Lemma 6. *The character* $\gamma_p\langle 4pu\rangle$ *is equal to* $\left(\dfrac{u}{p}\right)\gamma_p\langle 4p\rangle$.

Here the Legendre symbol (u/p) is defined to be either $+1$ or -1 according as u is or is not a quadratic residue modulo p.

Proof. Proceeding as above, the p-primary component $(L^\#/L)_p$ is spanned by a vector l_1/p with $\varphi(l_1/p)\equiv 2u/p\,(\text{mod}\ \mathbf{Z})$. Hence

$$\gamma_p\langle 4pu\rangle=\sum_{j=1}^{p}\exp\bigl(2\pi i(2u\,j^2/p)\bigr).$$

If $(u/p)=+1$, then evidently the expression $u\,j^2$ varies over all quadratic residues modulo p, taking each non-zero value twice, and taking the value zero modulo p just once. On the other hand if $(u/p)=-1$ then uj^2 takes each non-residue value twice, again taking the zero value once. Since the sum of $\exp(2\pi i(2k/p))$ over *all* residue classes k modulo p is zero, the conclusion follows easily. (Compare [Lang, p. 85].) □

Now let p and q be distinct odd primes. Applying the reciprocity formula

$$\gamma_2(L)\,\gamma_p(L)\,\gamma_q(L)\,\gamma_\infty(L)$$

to the lattice $L=\langle 4pq\rangle$ we obtain the identity

$$\left(\frac{p}{q}\right)\left(\frac{q}{p}\right)\exp(2\pi i(p-1)(q-1)/8)=1.$$

This is just the classical quadratic reciprocity law.

Concluding remark. There is an analogous Weil reciprocity formula over an arbitrary number field, which can be derived from the rational reciprocity formula. (Compare [Scharlau, 1972] and [Knebusch-Scharlau, 1971].) It takes the form

$$\prod_v \gamma_v(X)=1,$$

where X is an inner product space over the number field F, and v ranges over all valuations of F. Here γ_v is defined as follows.

Case 1. If v is a complex archimedian valuation, then $\gamma_v(X)=1$.

Case 2. If v is a real archimedian valuation, then

$$\gamma_v(X)=\exp(-2\pi i\,\sigma_v(X)/8)$$

where $\sigma_v(X)$ is the associated signature.

Case 3. If v is the $\mathfrak{p}$-adic valuation, where $\mathfrak{p}$ is a prime ideal in the ring of integers D, then $\gamma_v(X)$ is defined as the ratio

$$G((L^\#/L)_\mathfrak{p})/\sqrt{|(L^\#/L)_\mathfrak{p}|}.$$

Here L can be any D-lattice in X satisfying $\frac{1}{2}l\cdot l\in D$ for $l\in L$, and $L^\#$ is the dual lattice with respect to the $\mathbf{Q}$-valued inner product

$$x, y\mapsto \operatorname{trace}_{F/\mathbf{Q}}\, x\cdot y.$$

The Gauss sum $G((L^\#/L)_\mathfrak{p})$ is defined as the sum of $\exp(\pi i\operatorname{trace}(u\cdot u))$ over all u in $(L^\#/L)_\mathfrak{p}$.

Now the reciprocity law $\prod_v \gamma_v(X)=1$ follows easily from Milgram's formula, applied to the inner product $\operatorname{trace}(x\cdot y)$ on L.

One special case of this reciprocity law is of particular interest. Suppose that

$$X=(\langle\alpha\rangle\oplus\langle-1\rangle)\otimes(\langle\beta\rangle\oplus\langle-1\rangle),$$

representing an element in the ideal $I^2(F)$ in the Witt ring. Then it is easily verified that $\gamma_v(X)=\pm 1$. If $F_v \not\cong \mathbf{C}$, then both Weil and Scharlau show that $\gamma_v(X)=-1$ for suitably chosen α and β. The correspondence

$$\alpha, \beta \mapsto \gamma_v((\langle\alpha\rangle-\langle 1\rangle)(\langle\beta\rangle-\langle 1\rangle))=\pm 1$$

is in fact the "Hilbert symbol" associated with the valuation v. The equation $\prod_v \gamma_v((\langle\alpha\rangle-\langle 1\rangle)(\langle\beta\rangle-\langle 1\rangle))=1$ is Hilbert's form of the quadratic reciprocity law. (Compare [O'Meara], as well as Chapter III, § 5.4–5.9, and Chapter IV, § 4.4.)

Appendix 5. The Leech Lattice, and Other Lattices in Dimension 24

We will construct a self-dual unimodular lattice $L \subset \mathbf{R}^{24}$ with the property that $x \cdot x \geqq 4$ for every $x \neq 0$ in L. (Compare Chapter II, §§ 6, 7.)

The construction begins with the following combinatorial statement. Let $\mathbf{F}_2^{24}$ denote the vector space over $\mathbf{F}_2$ consisting of all 24-tuples of integers modulo 2.

Lemma. *There exists a* 12-*dimensional subspace* $S \subset \mathbf{F}_2^{24}$ *with the following property. For every non-zero vector* $s = (s_1, \ldots, s_{24})$ *in* S, *the number of components* s_i *which are equal to one is at least* 8, *and is divisible by* 4. *Furthermore* S *contains the vector* $(1, \ldots, 1)$ *consisting of* 24 *ones.*

Proof. Following Leech, we will display S as the row space of an explicit 12×24 matrix. Let A denote the symmetric 11×11 matrix over $\mathbf{F}_2$ whose first row is

$$1 \quad 1 \quad 1 \quad 0 \quad 1 \quad 1 \quad 0 \quad 1 \quad 0 \quad 0 \quad 0$$

and whose remaining rows are obtained by permuting these entries cyclically to the left. Thus each row of A contains 6 ones. Patient inspection shows that

(i) *each pair of distinct rows of* A *has precisely* 3 *ones in common* (i.e., in the same column).

Let B denote the symmetric 12×12 matrix

$$B = \left[\begin{array}{c|c} 0 & 1 \quad 1 \quad \ldots \quad 1 \\ \hline 1 & \\ 1 & \\ \vdots & A \\ 1 & \end{array}\right]$$

which is formed from A by adjoining the first row 0 1 1 1 1 1 1 1 1 1 1 1 and a corresponding first column. Using (i) we easily verify the following.

(ii) *The matrix B satisfies the identity* $B^2 = BB^t = I$. *Hence B is non-singular, and any two rows of B are orthogonal with respect to the inner product* $r \cdot r' = \sum r_i r'_i$.

The block matrix

$$C = \boxed{I \mid B}$$

is now the required 12×24 matrix of rank 12. Evidently the sum of all of the rows of C is equal to the 24-tuple $(1, 1, \ldots, 1)$. Note that

(iii) *The number of ones in any row of C is either* 8 *or* 12. *Furthermore any two distinct rows of C are orthogonal.*

It will be convenient to use the notation $\|s\|$ for the number of ones in a 24-tuple $s = (s_1, \ldots, s_{24})$. As a corollary of (iii) we obtain the following statement.

(iv) *If s is a linear combination of the rows of C, then* $\|s\| \equiv 0 \pmod 4$.

This is proved by induction on the number of rows involved. If s' is obtained from s by adding a row r, then evidently

$$\|s'\| = \|s\| + \|r\| - 2n$$

where n denotes the number of ones which s and r have in common. But s and r are orthogonal by (iii), so n is even. Assuming inductively that $\|s\|$ is divisible by 4, it follows that $\|s'\|$ is divisible by 4 also.

(v) *If s is a non-zero linear combination of the rows of C, then* $\|s\| \geqq 8$.

Proof. By (iv), it suffices to prove that $\|s\| \geqq 5$. Suppose that s is the sum of k distinct rows of C. The case $k=1$ is covered by (iii). If $k=2$, then it follows easily from (i) that $\|s\| = 8$. If $k=3$ and if s is the sum of the first row of C and two other rows, then again it follows from (i) that $\|s\| = 8$. If s is the sum of three rows of C not including the first row, then evidently the first thirteen entries of s include precisely 4 ones. If the remaining eleven entries were all zero, this would mean that the sum of the three corresponding rows of A was zero. Hence the sum of the remaining eight rows of A would also be zero; and the sum of the corresponding eight rows of B would be zero, contradicting (ii). Therefore $\|s\| \geqq 5$.

Finally, if $k \geqq 4$, then the first twelve entries of s contain at least 4 ones, and the remaining entries contain at least 1 one by (ii), so again it follows that $\|s\| \geqq 5$. This proves (v), and completes the proof of the Lemma. □

Remark 1. The matrix C was constructed in a rather ad hoc manner. The following description of its row space S may seem a little more motivated. Consider the field $\mathbf{F}_{2048}$ with 2^{11} elements. We claim that S

can be identified with the collection of all "relations" between the 23rd roots of unity in $\mathbf{F}_{2048}$. Let ω denote a 23rd root of unity satisfying the irreducible equation

$$1+\omega+\omega^5+\omega^6+\omega^7+\omega^9+\omega^{11}=0$$

over $\mathbf{F}_2$, and let φ denote the Frobenius automorphism $\alpha \mapsto \alpha^2$. Then the list

$$\varphi^{11}(\omega^{-1}),\ \varphi^{10}(\omega^{-1}),\ \ldots,\ \varphi^{1}(\omega^{-1}),\ 1,\ \varphi^{1}(\omega),\ \varphi^{2}(\omega),\ \ldots,\ \varphi^{11}(\omega)$$

contains each 23rd root of unity in $\mathbf{F}_{2048}$ just once. Now S is the collection of all 24-tuples $(s_1, \ldots, s_{24})$ of integers mod 2, with sum zero, satisfying the linear relation

$$s_2\,\varphi^{11}(\omega^{-1})+s_3\,\varphi^{10}(\omega^{-1})+\cdots+s_{12}\,\varphi^{1}(\omega^{-1})+s_{13}+s_{14}\,\varphi^{1}(\omega)$$
$$+\cdots+s_{24}\,\varphi^{11}(\omega)=0.$$

Details will be omitted.

Remark 2. There is an interesting simple group associated with the matrix C. The *Mathieu group* M_{24} can be defined as the group of all permutations of the columns of C which transform each row into a linear combination of the rows of C. This group acts 5-fold transitively, and has order $24 \cdot 23 \cdot 22 \cdot 21 \cdot 20 \cdot 48$. Although it was described by Mathieu in 1861, its existence was first firmly established by Witt in 1938 (Hamburg Abh. **12**).

We are now ready to construct the Leech lattice. Let L_0 denote the lattice in $\mathbf{R}^{24}$ with orthogonal basis $b_1, \ldots, b_{24}$ where $b_i \cdot b_i = \frac{1}{8}$. Let L denote the sublattice of L_0 consisting of all linear combinations $t_1 b_1 + \cdots + t_{24} b_{24}$ with integer coefficients such that either

(vi) *the integers* $t_1, \ldots, t_{24}$ *are all even, with* $t_1+\cdots+t_{24} \equiv 0$ (mod 8), *and the 24-tuple* $\frac{1}{2}(t_1, \ldots, t_{24})$ *reduced modulo 2 belongs to the vector space S of the lemma; or*

(vii) *the integers* $t_1, \ldots, t_{24}$ *are all odd, with* $t_1+\cdots+t_{24} \equiv 4$ (mod 8), *and the 24-tuple* $\frac{1}{2}(1+t_1, \ldots, 1+t_{24})$ *reduced modulo 2 belongs to S.*

An easy argument shows that L is closed under addition, forming a sublattice of index 2^{36} in L_0. Therefore

$$\det L = 4^{36} \det L_0 = 1.$$

We will show that the norm $\frac{1}{8}(t_1^2+\cdots+t_{24}^2)$ *of an element of L is always an even integer.* In other words

$$t_1^2+\cdots+t_{24}^2 \equiv 0 \pmod{16}.$$

If the t_i are even, then t_i^2 is congruent to 0 or 4 modulo 16 according as t_i is congruent to 0 or 2 modulo 4. But it follows from the lemma that the number of t_i congruent to 2 modulo 4 is divisible by 4. If the t_i are odd, then t_i^2 is congruent to 1 or 9 modulo 16 according as t_i is congruent to ± 1 or ± 3 modulo 8. Thus

$$\sum t_i^2 \equiv \alpha_1 + 9\alpha_3 + 9\alpha_5 + \alpha_7 \pmod{16},$$

where α_j denotes the number of t_i which are congruent to j modulo 8. Note the congruences

$$\begin{aligned} \alpha_1 + \alpha_3 + \alpha_5 + \alpha_7 = 24 &\equiv 0 \pmod{8}, \\ \alpha_1 + 3\alpha_3 + 5\alpha_5 + 7\alpha_7 &\equiv 4 \pmod{8}, \\ \alpha_1 \qquad + \alpha_5 \qquad &\equiv 0 \pmod{4}, \end{aligned}$$

where the last two follow from (vii) and the lemma. Adding the first two congruences and subtracting twice the third, we obtain

$$4\alpha_3 + 4\alpha_5 \equiv 4 \pmod{8}.$$

Thus $\alpha_3 + \alpha_5$ is odd, and it follows that $\sum t_i^2 \equiv 24 + 8(\alpha_3 + \alpha_5) \equiv 0 \pmod{16}$.

For any x and y in L, it follows that

$$x \cdot y = \tfrac{1}{2}((x+y)\cdot(x+y) - x \cdot x - y \cdot y) \in \mathbf{Z}.$$

Thus the lattice L is self-dual.

Note that no element $x \in L$ can satisfy $x \cdot x = 2$. For if $t_1^2 + \cdots + t_{24}^2 = 16$ then the t_i certainly cannot all be odd. But the only expressions for 16 as a sum of even squares are

$$16 = 4^2 = 2^2 + 2^2 + 2^2 + 2^2,$$

and both possibilities are excluded by (vi) and the lemma. Therefore $x \cdot x \geqq 4$ for every $x \neq 0$.

For further information about the Leech lattice, the reader is referred to [Conway].

Concluding remark. A complete classification of unimodular lattices of type II in $\mathbf{R}^{24}$ has been given by [Niemeier]. He shows that there are precisely 24 such lattices L up to isomorphism; and that a complete invariant for L is provided by the finite subset $R(L)$ consisting of all vectors $x \in L$ with norm $x \cdot x$ equal to 2. Evidently, for any $x_0 \in R(L)$ the reflection

$$y \mapsto y - (x_0 \cdot y)\, x_0$$

in the hyperplane perpendicular to x_0 maps L to itself, and hence maps $R(L)$ to itself. Therefore $R(L)$ is a "root system", as described in [Bourbaki, v. **34**, pp. 142–197], in some euclidean space. Note that the angle between any two vectors in $R(L)$ is either 0°, 60°, 90°, 120°, or 180°.

Using the classification theorem for root systems, we see that $R(L)$ is a disjoint union of mutually perpendicular root systems, each of which can be described by a "Dynkin diagram" of one of the following three types. In each case, each vertex of the Dynkin diagram represents a basis vector of norm $x \cdot x = 2$ in an m-dimensional lattice, and two such basis vectors have inner product either -1 or 0 according as they are joined by a line segment or not. The associated root system consists of *all* vectors of norm 2 in the lattice $\bar{L}$ spanned by these basis vectors.

Type $A_m (m \geqq 1)$. In this case the Dynkin diagram consists of m vertices joined by $m-1$ line segments as follows.

In terms of auxilliary orthonormal vectors $e_1, \ldots, e_{m+1}$, the i-th vertex in this diagram can be identified with the difference vector $e_i - e_{i+1}$. Thus the lattice $\bar{L}$ can be identified with the lattice consisting of all $(m+1)$-tuples of integers with sum zero. The determinant of $\bar{L}$ is equal to $m+1$.

Type $D_m (m \geqq 4)$. In this case the m vertices are connected as follows.

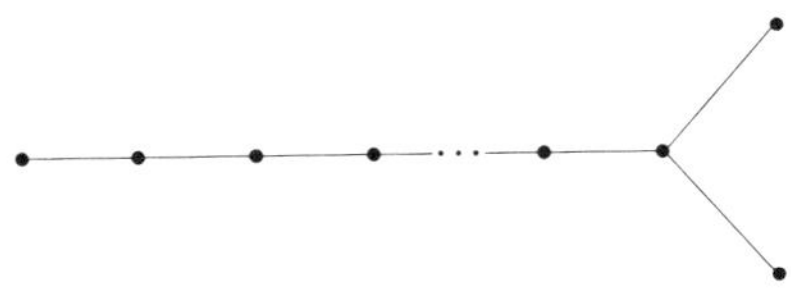

In terms of orthonormal vectors $e_1, \ldots, e_m$, the m vertices can be identified with the vectors $e_i - e_{i+1}$ and $e_{m-1} + e_m$. The lattice $\bar{L}$ can be identified with the lattice consisting of all m-tuples of integers with even sum. Its determinant is equal to 4.

Type $E_m (m = 6, 7, 8)$. In these three exceptional cases the m vertices are connected as follows; and the determinant of $\bar{L}$ is equal to $9-m$.

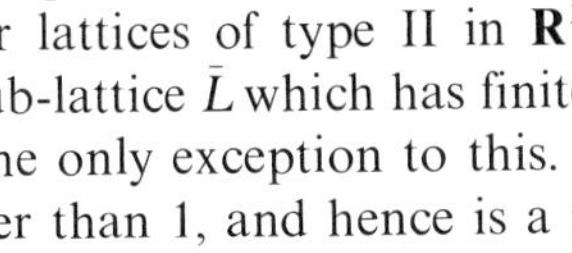

Compare the discussion in Chapter II, § 7.

Niemeier gives an explicit list of the 24 distinct root systems which arise from unimodular lattices of type II in $\mathbf{R}^{24}$. In general the root system $R(L)$ spans a sub-lattice $\bar{L}$ which has finite index in L. The Leech lattice, with $\bar{L} = 0$, is the only exception to this. In general the lattice $\bar{L}$ has determinant greater than 1, and hence is a proper sub-lattice of L. Again there is just one exception, namely the lattice $\bar{L} = L = \Gamma_8 \oplus \Gamma_8 \oplus \Gamma_8$ with root system $R(L)$ equal to $E_8 \cup E_8 \cup E_8$. Note that $\bar{L}$ may be (and usually is) decomposable, even when the unimodular lattice L is indecomposable.

Chronological Table

Adrien-Marie Legendre	1752–1833
Carl Friedrich Gauss	1777–1855
James Joseph Sylvester	1814–1897
Charles Hermite	1822–1901
Gotthold Eisenstein	1823–1852
Leopold Kronecker	1823–1891
Henry John Stephen Smith	1826–1883
Aleksandr Nikolaevič Korkin	1837–1908
Arnold Meyer	1844–1896
Egor Ivanovič Zolotarëv	1847–1878
Hermann Minkowski	1864–1909
Carl Ludwig Siegel	1896–
Helmut Hasse	1898–
Ernst Witt	1911–
Albrecht Pfister	1934–

References

Arason, J. K., Pfister, A.: Beweis des Krullschen Durchschnittsatzes für den Wittring. Invent. Math. **12**, 173–176 (1971).

Arf, C.: Untersuchungen über quadratische Formen in Körpern der Characteristic 2. J. Reine Angew. Math. **183**, 148–167 (1941). (Compare ibid. **193**, 121–125 (1954).)

Artin, E.: Geometric algebra. Interscience 1957.

Bass, H.: Lectures on topics in algebraic *K*-theory. Bombay: Tata Institute 1967.

Birkhoff, G., MacLane, S.: A survey of modern algebra. MacMillan 1941.

Blichfeldt, H. F.: The minimum values of quadratic forms in six, seven, and eight variables. Math. Z. **39**, 1–15 (1935).

Blij, F. van der: An invariant of quadratic forms mod 8. Indag. Math. **21**, 291–293 (1959).

Borevich, Z. I., Shafarevich, I. R.: Number theory. Academic Press 1966.

Bourbaki, N.: Éléments **24** (Algèbre 9), Formes sesquilinéaires et formes quadratiques. Hermann 1959.

Bourbaki, N.: Éléments **34**, Groupes et algèbres de Lie 4–6. Hermann 1968.

Braun, H.: Geschlechter quadratischer Formen. J. Reine Angew. Math. **182**, 32–49 (1940).

Cartan, H., Eilenberg, S.: Homological algebra. Princeton University Press 1956.

Chevally, C.: The algebraic theory of Spinors. Columbia University Press 1954.

Conway, J. H.: A group of order 8,315,553,613,086,720,000. Bull. Lond. Math. Soc. **1**, 79–88 (1969).

Conway, J. H.: A characterization of Leech's lattice. Invent. Math. **7**, 137–142 (1969).

Conway, J. H.: Groups, lattices, and quadratic forms, pp. 135–139 of "Computers in Algebra and Number Theory", SIAM-AMS Proceedings **4**, AMS 1971.

Dickson, L. E.: History of the Theory of Numbers, **2** and **3**, New York: G. E. Stechart & Co. 1934.

Dyson, F. J.: Quaternion determinants. Helv. Phys. Acta **45**, 289–302 (1972).

Fröhlich, A.: Discriminants of algebraic number fields. Math. Z. **74**, 18–28 (1960). (See also: Ideals in an extension field ..., pp. 29–38.)

Fröhlich, A.: Hermitian and quadratic forms over rings with involution. Quart. J. Math. Oxford **20**, 297–317 (1969).

Fröhlich, A.: On the *K*-theory of unimodular forms over rings of algebraic integers. Quart. J. Math. Oxford, to appear.

Fröhlich, A., McEvett: Forms over rings with involution. J. Algebra **12**, 79–104 (1969).

Gel'fand, I. M., Mischchenko, A. S.: Quadratic forms over commutative group rings and the *K*-theory. Functional Analysis and its Applications **3**, 277–281 (1969).

Geyer, W.-D., Harder, G., Knebusch, M., Scharlau, W.: Ein Residuensatz für symmetrische Bilinearformen. Invent. Math. **11**, 319–328 (1970).

Hilbert, D., Cohn-Vossen, S.: Geometry and the imagination. Chelsea 1952.

Hirzebruch, F.: Topological methods in algebraic geometry. Springer 1966.

Jacobson, N.: A note on hermitian forms. Bull. Amer. Math. Soc. **46**, 264–268 (1940).

Knebusch, M.: Grothendieck und Wittringe von nichtausgearteten symmetrischen Bilinearformen, Sitzungsber. Heidelb. Akad. Wiss. Math.-naturw. Kl. 1969/70, 3. Abh.

Knebusch, M.: Runde Formen über semilokalen Ringen. Math. Ann. **193**, 21–34 (1971).
Knebusch, M.: Ein Satz über die Werte von quadratische Formen über Körpern. Invent. Math. **12**, 300–303 (1971).
Knebusch, M., Rosenberg, A., Ware, R.: Structure of Witt rings, quotients of abelian group rings, and orderings of fields. Bull. Amer. Math. Soc. **77**, 205–209 (1971).
Knebusch, M., Rosenberg, A., Ware, R.: Structure of Witt rings and quotients of Abelian group rings. Amer. J. Math. **44**, 119–155 (1972).
Knebusch, M., Rosenberg, A., Ware, R.: Grothendieck and Witt rings of hermitian forms over Dedekind rings. Pacific J., to appear.
Knebusch, M., Rosenberg, A., Ware, R.: Signatures on semilocal rings. Bull. Amer. Math. Soc. **78**, 62–64 (1972).
Knebusch, M., Scharlau, W.: Über das Verhalten der Witt-Gruppe bei Körpererweiterungen. Math. Ann. **193**, 189–196 (1971).
Knebusch, M., Scharlau, W.: Quadratische Formen und quadratische Reziprozitätsgesetze über algebraischen Zahlkörpern. Math. Z. **121**, 346–368 (1971).
Kneser, M.: Zur Theorie der Kristallgitter. Math. Ann. **127**, 105–106 (1954).
Kneser, M.: Klassenzahlen definiter quadratischer Formen, Arch. Math. **8**, 241–250 (1957).
Korkine, A., Zolotareff, G.: Sur les formes quadratiques positives. Math. Ann. **11**, 242–292 (1877).
Landherr, W.: Äquivalenz Hermitescher Formen über einen beliebigen algebraischen Zahlkörper. Abh. Math. Sem. Univ. Hamburg **11**, 245–248 (1935).
Lang, S.: Algebraic number theory. Addison-Wesley 1970.
Leech, J.: Notes on sphere packings. Canad. J. Math. **19**, 251–267 (1967).
Leech, J., Sloane, N. J. A.: New sphere packings in dimensions 9–15. Bull. Amer. Math. Soc. **76**, 1006–1010 (1970).
Lorenz, F.: Quadratische Formen über Körpern. Lecture Notes in Mathematics **130**, Springer 1970.
Lorenz, F., Leicht, J.: Die Primideale des Wittschen Ringes. Invent. Math. **10**, 82–88
MacKenzie, R., Scheuneman, J.: A number field without a relative integral basis. Amer. Math. Monthly **78**, 882 (1971).
MacLane, S.: Hamiltonian mechanics and geometry. Amer. Math. Monthly **77**, 570–585 (1970).
Mars, J. G. M.: The Siegel formula for orthogonal groups, Proceedings Symposia in Pure Math. **9**, A. M. S. 1966, pp. 133–142.
Massey, W.: Algebraic topology. Harcourt, Brace and World, 1967.
Meyer, A.: Über die Auflösung der Gleichung $ax^2+by^2+cz^2+du^2+ev^2=0$ in ganzen Zahlen. Vierteljahrs. Naturfors. Gesells. Zürich **29**, 220–222 (1884).
Milnor, J.: On simply connected 4-manifolds, Symposium Internacional Topologia Algebraica, Mexico, 1958.
Milnor, J.: On isometries of inner product spaces. Invent. Math. **8**, 83–97 (1969).
Milnor, J.: Algebraic K-theory and quadratic forms. Invent. Math. **9**, 318–344 (1970).
Milnor, J.: Symmetric inner products in characteristic 2, in Prospects in Mathematics, Annals Study **70**, Princeton University Press 1971.
Milnor, J.: Introduction to algebraic K-theory. Annals Studies **72**, Princeton University Press 1971.
Niemeier, H.-V.: Definite quadratische Formen der Diskriminante 1 und Dimension 24. J. Number Theory (to appear).
O'Meara, O. T.: Introduction to quadratic forms. Springer, 1963.
Pfister, A.: Quadratische Formen in beliebigen Körpern. Invent. Math. **1**, 116–132 (1966).
Rogers, C. A.: The packing of equal spheres. Proc. London Math. Soc. **8**, 609–620 (1958).
Rogers, C. A.: Packing and covering. Cambridge University Press 1964.
Sah, Chih-Han: Symmetric bilinear forms and quadratic forms. J. Algebra **20**, 144–160 (1972).

Scharlau, W.: Quadratic forms, Queen's papers on pure and applied math. 22, Queen's Univ., Kingston, Ontario, 1969.

Scharlau, W.: Zur Pfisterschen Theorie der quadratischen Formen. Invent. Math. **6**, 327–328 (1969).

Scharlau, W.: Induction theorems and the structure of the Witt group, Invent. Math. **11**, 37–44 (1970).

Scharlau, W.: Hermitesche Formen über lokalen Körpern. Math. Ann. **186**, 201–208 (1970).

Scharlau, W.: Quadratic reciprocity laws. J. Number Theory **4**, 78–97 (1972).

Serre, J.-P.: Cours d'arithmétique, Presses Univ. France, 1970. (English translation, to appear.)

Siegel, C. L.: Gesammelte Abhandlungen I. Springer 1966, pp. 326–405.

Spanier, E.: Algebraic topology. McGraw-Hill 1966.

Springer, T. A.: Quadratic forms over fields with a discrete valuation I, Indag. Math. **17**, 352–362 (1955).

Steenrod, N.: The topology of fibre bundles. Princeton University Press 1951.

Swan, R.: Vector bundles and projective modules. Trans. Amer. Math. Soc. **105**, 264–277 (1962).

Swan, R.: Algebraic *K*-theory, Lecture Notes in Mathematics **76**, Springer 1968.

Waerden, B. L. van der: Die Reduktionstheorie der positiven quadratischen Formen, Acta Math. **96**, 265–309 (1956).

Wall, C. T. C.: Surgery on compact manifolds, Academic Press 1970.

Watson, G. L.: Integral quadratic forms. Cambridge University Press 1960.

Weil, A.: Sur certains groupes d'operateurs unitaires, Acta Math. **111**, 143–211 (1964).

Weil, A.: Sur la formule de Siegel dans la théorie des groupes classiques. Acta Math. **113**, 1–87 (1965).

Witt, E.: Theorie der quadratischen Formen in beliebigen Körpern. J. Reine Angew. Math. **176**, 31–44 (1937).

Zariski, O., Samuel, P.: Commutative Algebra **1**, Van Nostrand 1958.

Index

Special Notations

Ergebnisse der Mathematik und ihrer Grenzgebiete

1. Bachmann: Transfinite Zahlen. DM 48,—; US $15.30
2. Miranda: Partial Differential Equations of Elliptic Type. DM 58,—; US $18.40
4. Samuel: Méthodes d'Algèbre Abstraite en Géométrie Algébrique. DM 34,—; US $10.80
5. Dieudonné: La Géométrie des Groupes Classiques. DM 42,—; US $13.40
7. Ostmann: Additive Zahlentheorie. 1. Teil: Allgemeine Untersuchungen. DM 42,—; US $13.40
8. Wittich: Neuere Untersuchungen über eindeutige analytische Funktionen. DM 36,—; US $13.40
11. Ostmann: Additive Zahlentheorie. 2. Teil: Spezielle Zahlenmengen. DM 34,—; US $10.80
13. Segre: Some Properties of Differentiable Varieties and Transformations. DM 46,—; US $14.60
14. Coxeter/Moser: Generators and Relations for Discrete Groups. DM 42,—; US $13.40
15. Zeller/Beckmann: Theorie der Limitierungsverfahren. DM 64,—; US $20.30
16. Cesari: Asymptotic Behavior and Stability Problems in Ordinary Differential Equations. DM 54,—; US $17.20
17. Severi: Il theorema di Riemann-Roch per curve – superficie e varietà questioni collegate. DM 30,—; US $9.60
18. Jenkins: Univalent Functions and Conformal Mapping. DM 37,—; US $11.80
19. Boas/Buck: Polynomial Expansions of Analytic Functions. DM 24,—; US $7.70
20. Bruck: A Survey of Binary Systems. DM 46,—; US $14.60
21. Day: Normed Linear Spaces. In preparation
23. Bergmann: Integral Operators in the Theory of Linear Partial Differential Equations. DM 40,—; US $12.70
25. Sikorski: Boolean Algebras. DM 42,—; US $13.40
26. Künzi: Quasikonforme Abbildungen. DM 43,—; US $13.70
27. Schatten: Norm Ideals of Completely Continuous Operators. DM 30,—; US $9.60
28. Noshiro: Cluster Sets. DM 40,—; US $12.70
30. Beckenbach/Bellman: Inequalities. DM 38,—; US $12.10
31. Wolfowitz: Coding Theorems of Information Theory. DM 30,—; US $9.60
32. Constantinescu/Cornea: Ideale Ränder Riemannscher Flächen. DM 75,—; US $23.80
33. Conner/Floyd: Differentiable Periodic Maps. DM 34,—; US $10.80
34. Mumford: Geometric Invariant Theory. DM 24,—; US $7.70
35. Gabriel/Zisman: Calculus of Fractions and Homotopy Theory. DM 42,—; US $13.40
36. Putnam: Commutation Properties of Hilbert Space Operators and Related Topics. DM 31,—; US $9.90
37. Neumann: Varieties of Groups. DM 51,—; US $16.20
38. Boas: Integrability Theorems for Trigonometric Transforms. DM 20,—; US $6.20
39. Sz.-Nagy: Spektraldarstellung linearer Transformationen des Hilbertschen Raumes. DM 24,—; US $7.70
40. Seligman: Modular Lie Algebras. DM 43,—; US $13.70
41. Deuring: Algebren. DM 30,—; US $9.60
42. Schütte: Vollständige Systeme modaler und intuitionistischer Logik. DM 30,—; US $9.60
43. Smullyan: First-Order Logic. DM 36,—; US $11.50
44. Dembowski: Finite Geometries. DM 68,—; US $21.60
45. Linnik: Ergodic Properties of Algebraic Fields. DM 44,—; US $14.00
46. Krull: Idealtheorie. DM 34,—; US $10.80
47. Nachbin: Topology on Spaces of Holomorphic Mappings. DM 18,—; US $5.80

48. A. Ionescu Tulcea/C. Ionescu Tulcea: Topics in the Theory of Lifting. DM 36,—; US $11.50
49. Hayes/Pauc: Derivation and Martingales. DM 48,—; US $15.30
50. Kahane: Séries de Fourier absolument convergentes. DM 44,—; US $14.00
51. Behnke/Thullen: Theorie der Funktionen mehrerer komplexer Veränderlichen. DM 48,—; US $15.30
52. Wilf: Finite Sections of Some Classical Inequalities. DM 28,—; US $8.90
53. Ramis: Sous-ensembles analytiques d'une variété banachique complexe. DM 36,—; US $11.50
54. Busemann: Recent Synthetic Differential Geometry. DM 32,—; US $10.20
55. Walter: Differential and Integral Inequalities. DM 74,—; US $23.50
56. Monna: Analyse non-archimédienne. DM 38,—; US $12.10
57. Alfsen: Compact Convex Sets and Boundary Integrals. DM 46,—; US $14.60
58. Greco/Salmon: Topics in m-Adic Topologies. DM 24,—; US $7.70
59. López de Medrano: Involutions on Manifolds. DM 36,—; US $11.50
60. Sakai: C*-Algebras and W*-Algebras. DM 68,—; US $21.60
61. Zariski: Algebraic Surfaces. DM 54,—; US $17.20
62. Robinson: Finiteness Conditions and Generalized Soluble Groups, Part 1. DM 48,—; US $15.30
63. Robinson: Finiteness Conditions and Generalized Soluble Groups, Part 2. DM 64,—; US $20.30
64. Hakim: Topos annelés et schémas relatifs. DM 48,—; US $15.30
65. Browder: Surgery on Simply-Connected Manifolds. DM 42,—; US $13.40
66. Pietsch: Nuclear Locally Convex Spaces. DM 48,—; US $15.30
67. Dellacherie: Capacités et processus stochastiques. DM 44,—; US $14.00
68. Raghunathan: Discrete Subgroups of Lie Groups. DM 56,—; US $17.80
69. Rourke/Sanderson: Introduction to Piecewise-Linear Topology. DM 42,—; US $13.40
70. Kobayashi: Transformation Groups in Differential Geometry. DM 52,—; US $16.50
71. Tougeron: Idéaux de fonctions différentiables. DM 69,—; US $21.90
72. Gihman/Skorohod: Stochastic Differential Equations. DM 88,—; US $27.90
73. Milnor/Husemoller: Symmetric Bilinear Forms. DM 42,—; US $13.40
74. Fossum: The Divisor Class Group of a Krull Domain. DM 44,—; US $14.00